T0364514

SASHA FENTON

THE MAGIC OF ASTROLOGY

FOR HEALTH, HOME AND HAPPINESS

COLLINS & BROWN

INTRODUCTION

Are you an emotional Scorpio who needs to pour out your latest drama to your friends and relatives? Perhaps you are an uptight Capricorn who keeps all your feelings to yourself. Are you a flirtatious Libran, a brave but thoughtless Arian or a careful Cancerian? Do your home and garden display Taurean flair or are your plants neatly laid out in Virgoan rows? Are you a Leo who earns a frequent-user discount while online shopping or at the beauty salon? Or a Sagittarian quietly and desperately yearning for the wide open spaces while surrounded by a nagging family and a household full of pets? Do your poor, painful Piscean feet make a martyr of you? Are you an Aquarian who is permanently searching for more and more knowledge? Whatever your individual motivations and peculiarities might be, you should be able to find them in this book.

SUN SIGN ASTROLOGY

Your sun sign or star sign is simply the sign of the zodiac that you were born under. With a good understanding of sun sign astrology, the reasons behind your behaviour patterns – and even some aspects of your past and future – can become clearer and can help you to understand the choices you find yourself drawn to in life. If you want to know why you prefer your home to be minimalist and spartan, or why you like it crammed full of dust-gathering knick-knacks, your sun sign could hold the answer. If, however, you are not typical of your sun sign, it is worth reading about other sun signs in order to discover which one could be exerting a strong influence over the decisions you make every day. The beauty of sun sign astrology is in its simplicity: all you need to know to start is your date of birth. With this book, you can extend your knowledge to discovering your moon sign, your rising sign and all your other modifying features. But learning about the different ways in which a sun sign can influence every aspect of your life – from the partners you choose to the working role that brings out the best in you – makes an ideal starting point.

It is not difficult to work out which sun sign you belong to; even people who are on the cusp of two signs usually identify more with one than the other. As you read this book, you will find out in detail how your sun sign, and those of your loved ones, affects your motivations and your character... and even, perhaps, to discover what lies in store for you in the years ahead.

YOUR SUN SIGN DATES

Aries
March 20 – April 19

Taurus
April 20 – May 20

Gemini
May 21 – June 20

Cancer
June 21 – July 22

Leo
July 23 – August 22

Virgo
August 23 – September 22

Libra
September 23 – October 22

Scorpio
October 23 – November 21

Sagittarius
November 22 – December 21

Capricorn
December 22 – January 20

Aquarius
January 21 – February 19

Pisces
February 20 – March 19

Before the advent of telescopes, astronomers thought that the Sun travelled around the Earth, passing in front of each of the 12 constellations that lay along its path. We now know that the planets of the solar system orbit the Sun and only appear to be moving through these constellations from our perspective. The belt of the heavens encircling the Earth along the ecliptic (the apparent pathway of the Sun) is known as the zodiac, and it is divided into 12 equal segments: the constellations that are the signs of the zodiac. For example, on January 1 the sun appears to be travelling through the sign of Capricorn, while on July 1 it appears to be travelling through the sign of Cancer.

If you are unsure of your sun sign, refer to the table on this page. To complicate matters, unfortunately each sun sign changes at a slightly different time of day every year. Therefore, if you are born on the cusp of a sign, you may find that some astrologers give one date for the start or the end of your sign, while others go for a slightly different one.

The signs of the zodiac are always arrayed in the same order. They start with Aries, setting the astrological year rolling at the spring equinox on March 21, which the ancients saw as the beginning of the new year. Aries is always followed in turn by the sun signs Taurus, Gemini, Cancer, Leo, Virgo, Libra, Scorpio, Sagittarius, Capricorn, Aquarius and Pisces.

THE ELEMENTS

Fire, earth, air and water are the four elements from which, traditionally, the Earth and all who live on it are made. Each sign of the zodiac has an ancient link with one of these elements.

FIRE

Aries, Leo, Sagittarius

These signs are associated with initiative, courage and leadership. Fire sign people won't hesitate to plunge headlong into life, though can't always understand why others don't.

Fire people are quick, intelligent and generous and love friends, family and lovers with great passion. More sensitive than they appear, they can be cut to the quick when the affection and generosity they unstintingly give is not appreciated or returned in the same measure. Fire sign subjects cannot tolerate going without the best things in life and are prepared to work for them. They can be arrogant with those they consider inferior, and can make others feel inadequate.

EARTH

Taurus, Virgo, Capricorn

Practical, diligent and hard working, earth people are often happiest when doing something useful. They can be relied upon; though it takes them a while to get around to things, they usually do as they say they will. Shrewd and cautious, these people need material and emotional security and put up with a lot to get it. Many appear stingy, often driven by a fear of poverty.

The 'earthiness' of earth folk shows itself in their sensuality, and they also have a creative streak. Shy and sensitive, earth signs cling to their families and keep outsiders at a distance, but once they make a friend they are loyal, unless the friend takes advantage financially. When they fall deeply in love, earth types are dependable, but it can take them half a lifetime to find a partner who is just right.

AIR

Gemini, Libra, Aquarius

The intellectual group, with the mind taking precedence over physical activity. If an idea is needed, look to air sign people.

Gemini and Libra are sociable, often dealing with others as part of their work. Aquarians are great at deep discussions, but need time alone to delve into their inner world, and can occasionally lose touch with reality. Some air sign folk are happier dealing with computers and gadgets than with people, but most love company. They are excellent communicators, though may find it hard to empathize with others.

Restless and nervous, air signs need an outlet for their tensions. Geminis worry about trivia, while Librans can get very confused. Aquarians may become extremely hostile when things don't turn out their way.

WATER

Cancer, Scorpio, Pisces

Water people need time to grasp new situations or ideas because these need to be filtered through their feelings before they can assess them. They can be very emotional, but they keep their feelings hidden, sometimes even from themselves, which allows resentment to build up that can be expressed in inappropriate anger. Their loved ones live alongside their spells of boundless excitement, or terrible depressions. Water people attach themselves strongly to those they love.

Water folk are extremely intuitive, feeling everything in the atmosphere around them, and use this knowledge to smooth their pathway through life. Water signs make use of intuition in business, although Pisceans are less ambitious and can miss out on good opportunities.

Water subjects don't approach matters directly, but test out a situation and see if there is a way they can manoeuvre themselves into, out of, or around things.

THE QUALITIES

Every sun sign belongs to one of three qualities: cardinal, fixed or mutable. Although each sign also dwells in one of the four elements (see pages 5–7), it expresses itself differently according to its quality.

CARDINAL

Aries, Cancer, Libra, Capricorn

Cardinal signs see themselves as decision makers and won't allow their lives to be run by anyone else. This may not appear obvious in the case of Cancer and Libra, who can seem happy to let others take the lead. However, that is often simply a means of getting what they want. For example, a Cancerian can pretend to be weak, encouraging others to take on problems for them... it's just their way of getting the job done. This may look different from Arian aggression or Capricornian obstinacy, but it comes to the same thing in the end. Cardinality suggests self-motivation and self-interest. Librans can be good team leaders or business partners, as long as other aspects of their birth charts keep their feet on the ground. Business or group endeavours are important to Capricorns, but under their controlled exterior beats a loving heart and great family loyalty.

It is the cardinal person's desire to do just what they want in exactly the way they want. Their intentions may be admirable, but woe betide anyone who seeks to control or restrict them. Their own needs come first.

FIXED

Taurus, Leo, Scorpio, Aquarius

The key characteristic linking these very different signs is stubbornness and resistance to change. Fixed sign types try to stay in the same job for as long as possible, or will do their best to keep a good relationship going.

The most fixed is Taurus, which can be an obstinate sign. Aquarians stick to their opinions – unless they see a good reason to change them – while little shifts Leos from their objectives. With regard to their choice of career or partner, Scorpios appear more adaptable, but their loyalty and refusal to give way in the face of opposition is formidable. These signs are all reliable, though Aquarians can only be relied upon to do the unexpected!

Change unnerves these folk. If their way of life alters, they become downhearted and can lose the ability to function. Leos and Taureans can be extremely materialistic, marrying for money or staying in a miserable job for financial security. Scorpios are great collectors of eclectic junk that they cannot be parted from. Aquarians are the least materialistic of the group.

MUTABLE

Gemini, Virgo, Sagittarius, Pisces

The word mutable means something that can be changed, suggesting flexibility or adaptability. It is said that the cardinal signs set situations into motion, the fixed signs see things through and the mutable signs guide them through changes and bring things to a close.

Mutable sign people can adapt to change better than most, though they don't go looking for it. Geminis and Sagittarians need careers and lifestyles with plenty of variety. Pisceans and Virgoans can tolerate routine jobs as long as they have lots of outside interests for stimulation.

All mutable signs need creative outlets. Geminis may work in the field of communications or accountancy, but they aspire to businesses of their own, or work hard to make their homes perfect. Virgoans love to write and paint, while Sagittarians and Pisceans use their creativity in an artistic or spiritual manner, or by setting up organizations that work for the benefit of others.

The mutable signs know that change is inevitable and that, when it comes along, they must go with the flow.

RISING SIGNS

In addition to a sun sign, everyone has a rising sign (also called ascendant sign). To understand these two signs, remember that the Earth travels around the Sun every 365 days. But, from our point of view, the Sun appears to 'move', entering a different zodiac sign each month.

In addition to moving around the Sun, the Earth does a complete turn on its own axis every day. During the course of each day, therefore, the position of the sunrise moves gradually around the world. The combination of the date, the time of day and the place where you were born fixes the exact point where the Sun was rising at the moment of your birth. The sign that was coming up over the eastern horizon at the moment you were born is called your rising sign.

When people are untypical of their sun sign, it is often because of the rising sign overshadowing them. You can identify your rising sign through one of the many calculators online.

ARIES RISING

Outwardly serious, reserved and lacking in humour. Sport or physical work may figure strongly in your life. Childhood disagreements with your father were fierce.

TAURUS RISING

Pleasant and calm, but there is something seething underneath. Material gain is very important to you, and your attitude to money goes directly back to childhood experiences.

GEMINI RISING

Friendly, clever and versatile. An effective communicator, you talk with your face, hands and body as well as in the usual manner! Your childhood may have been difficult, so you lack confidence, but hide it well.

CANCER RISING

You shine in social situations. Although you can be a flirt, this is rarely acted on. Your childhood was fairly easy and materially comfortable. You are close to your mother and take responsibility for younger siblings.

LEO RISING

You love to be the centre of attention and are kind and generous. Your childhood was probably an extremely easy time in which your parents made you feel special.

VIRGO RISING

Loved for what you achieved as a child, you lose yourself in work as an adult, often working to serve the needs of others. You are modest and quiet but have a quick mind and great sense of humour.

LIBRA RISING

You get through life on your charm. You were either well loved and understood as a child, or somewhat neglected and given goods and outings instead of attention.

SCORPIO RISING

You find it hard to trust anybody, but, if you can trust a lover, you can build a good relationship. You may have been a secretive child and have learned to keep your own counsel.

SAGITTARIUS RISING

Friendly and intelligent, you are interested in people but treasure independence. Your parents may have come from a different background to that in which you grew up, so you may choose your values and ideals later in life.

CAPRICORN RISING

Ambitious and hardworking, you are likely to have had a difficult background. Your parents may have been too busy to spend time with you, throwing you back on your own resources.

AQUARIUS RISING

Outgoing and slightly eccentric, your childhood was probably filled with books and music and your success in life is either in technology or in a more unusual field, such as astrology.

PISCES RISING

You have a quiet nature but are defensive. As a child you may have avoided other children, preferring the world of books and the imagination. This may lead to an interest in mysticism and art in later life.

ARIES

YOUR SUN SIGN

Aries is a masculine fire sign for which the symbol is the ram. You have courage, a love of adventure, energy, initiative and a determination to live life to the full. You may look laid back, but you are extremely competitive. You play to win, and while this brings success, it can cost you true friends.

You love your home whether it is a dump or a palace, and resist interference in your domestic life. The chances are that you are competent at DIY but may go about it too quickly, without enough preparation, grabbing whatever tool is handy, even if it is someone else's best kitchen knife!

Despite being somewhat impulsive, you don't take chances where it matters, and can stay in an uninspiring job if you need the money. You tend to work best in a large organization, and are likely to wind up in a job requiring a hard hat.

You need an outlet for your competitiveness, and a sport could be just the thing. Many Arians enjoy following sport even if they don't play much themselves. You are sociable, with a great store of jokes and stories, and fond of a meal out and a drink or two. It's likely that you are an excellent singer or a good dancer, so you are always welcome at parties. You can't take too much sitting around indoors and you need something to look forward to at weekends. Although you are happy to spend on going out, you can be stingy about domestic expenses you find boring.

You often leap before looking, and this can land you in sticky situations. You may propose marriage while in a fit of romantic lust, only to find that your amour has feet of clay. Try to take more care and think before you speak.

The red planet, Mars, named after the Roman god of war, is associated with your courageous and assertive sign.

RULED BY MARS	YOUR ELEMENT
March 20 – April 19	Fire

FAMILY

THE FATHER

Always eager to educate his children, the Arian father reads to them, takes them on trips and exposes them to art and music. He can push his children too hard or tyrannize sensitive ones, but respects children who are high achievers and who stand up to him.

THE MOTHER

Though she loves her children dearly and will make amazing sacrifices for them, don't expect an Arian to give up her job or outside interests for motherhood. This mother wants her children to be the best and may push them too hard. She must remember to be affectionate to them.

THE CHILD

Lively, noisy and demanding, the Arian child is hard to ignore. They enjoy every moment of their childhood. Despite this, they are low in confidence and need reassurance. Often clever, but lacking self-discipline, they sometimes have to be made to attend school each day and do their homework.

FRIENDS

You find many kinds of friendship extremely valuable. Lesser friendships include those with whom you play sport, or meet for a drink and a chat at regular intervals. You form easy relationships with work colleagues, and these may spill over from working hours into time outside of work. Such friendships don't cost much in the way of time or emotional commitment, but they make your particular world go round.

You are so open and so lacking in suspicion or caution that you may expect too much of your friends. This makes it hard for you to accept that they can let you down from time to time. Remember that your pals may not think exactly as you do, but they are entitled to their opinions.

LOVE

Foods
Hot and spicy foods, onions, radishes, garlic

Indulgences
Alcohol and shopping

Exercise
Competitive sport, team games, squash

Colours
Crimson or carmine

Interests
Sport, nightclubs, drinking

Family activities
Outdoor activities and team games

Flowers
Honeysuckle

Holiday
Roaming the Greek islands

EMOTIONAL PATTERNS

If you are a male Arian, sexual availability and willingness attracts you. If female, you will find someone who seems happy to be led, as well as sexually attractive. Arians of both sexes want younger partners, as lovers of your own age seem harder to influence. You like treating your lover to meals out and trips, but can often be stingy. In short, your partner should be a companion and playmate, and not someone who needs long-term financial, emotional or physical support.

MARRIAGE AND PARTNERSHIPS

Many Arians marry young, and you may remain happily with your soul mate for life. You may be so close that you resent outsiders. If the marriage doesn't give you all that you want, you may stay with your partner, but complain until they are alongside you only in body, not soul. Some Arians are too choosy to make real relationships. Others have a multitude of exciting sexual encounters that never really mean anything.

LOVE AFFAIRS

Affairs are meat and drink to you. Arian men may lack confidence to go after someone until they are mature in years. Arian women look for a partnership, often in the wrong place, then become distraught when it doesn't work out. Many carry on an affair for years in the vain hope their lover will leave their spouse for them.

COMPATIBILITY

It is easy for Arians to become friends with fire signs: Aries, Leo and Sagittarius. All three are extrovert, proud and competitive. However, the similarities that make you good friends mitigate against happy marriage.

As regards feminine earth signs, Taurus might fit the bill if you need a practical partner, but nervous, nit-picking Virgo would drive you crazy. Determined, critical Capricorn shares your cardinal quality: you both like to get your own way.

Water signs – Cancer, Scorpio and Pisces – are least compatible with you due to their intuitive nature. You won't understand their approach to life or what they want from you.

Your best choice is an air sign. Although these are masculine, extrovert signs, their extroversion is less obvious than that of fire signs, so competition between you would be less.

WINNING COMBINATIONS

Aries with Gemini and Leo

WORK

WORKING ENVIRONMENT

Your sign is very competitive in all areas of life, but especially at work, and you thrive on being surrounded by a large number of people. Public service is likely to attract you, especially if it helps those who find themselves in emergency situations, which is why there are a lot of Arian police officers and paramedics. A job that carries an aspect of teaching, training or looking after inexperienced team members works well for you. Many Arians find their way into the teaching professions, the armed forces and the government, where there are many people around to get to know.

Your nature is intensely political and, if you don't actually find a committee or a political pressure group to join, you put your instinctive talent for office politics to use in some other way. An area of work such as the automobile industry or engineering is another option, because of your natural grasp of all things mechanical. You don't mind getting your hands dirty.

Whatever you do, you quickly work your way up the ladder of success through a mixture of intelligence and your ability to play the game of corporate politics better than most. Some Arians work very hard indeed, but other members of your sign can be lazy. Either way, once you find something that really grabs your interest, you make a success out of it through sheer hard work, a bit of luck, political adroitness and good judgement. When you reach a position of power, you must try to avoid bossiness or insensitive behaviour to those whose jobs are under your control.

YOUR ROLE AT WORK

Your choice of job involves either middle to upper management in a large organization, or running your own show completely. A typical example is a small crash-repair garage, where you not only run the business but can also get your hands dirty.

JOBS

Police officer, teacher, engineer

SPENDING MONEY ON...

Socializing, sport, friends

SAVING FOR...

Your children's education

TAURUS

YOUR SUN SIGN

Taurus is a feminine earth sign and the symbol is the bull. If you are a Taurus you are patient, thorough and reliable. You don't like change and you don't swap jobs or partners with ease, but you can cope with unexpected events better than most. Intensely loyal to your family, you love to be with them as much as possible, but if your children are grown-up, you don't cling to them. The most important thing for you is to be happy and comfortable in your home, surrounded by family, eating and entertaining as a loving, happy group.

As a child, you may have been a late developer due to your disinclination to take risks. You are shrewd, and read and know far more than people outside your family realize. Outsiders may not credit you with an exceptional mind, but you are often more logical and more likely to be correct than your friends.

You have a terrific eye for what looks attractive. Whether you build a home, create a garden or bake a cake, it will look good and fulfil its function perfectly. In addition, you have 'safe' hands, rarely dropping or breaking things; you keep your possessions, which you value a great deal, in good condition for many years. Car accidents are a rarity.

In business you can be ambitious, but have no interest in get-rich-quick schemes. You prefer a steady job, climbing the ladder of promotion slowly but surely. If you are in banking or finance, you can rise to a high level, but you may prefer something practical, such as the building trade, or something artistic, like a make-up artist or interior designer, or perhaps even someone in the confectionery trade.

There are some colourful rogue Taureans who dress outrageously, delight in bucking trends and frightening the establishment... or terrifying their in-laws. These types have a

RULED BY VENUS April 20 – May 20	YOUR ELEMENT Earth

wonderful sense of humour and love to upset everyone else's perception of their sign.

Venus, the Roman goddess of love, represents your sign and symbolizes the abundance and beauty of nature.

FAMILY

THE FATHER

Though he cares deeply for his children and wants the best for them, the Taurean father doesn't expect the impossible. He believes in discipline and may lay down the law. In fact, he has difficulty bridging the generation gap and can be unsympathetic to the attitudes and interests of young people.

THE MOTHER

A real earth mother, the Taurean woman enjoys baking bread and making wonderful toys and games for her children. Sane and sensible but not highly imaginative, she is best with a child who has conventional needs, but can be confused by one who is out of the ordinary in some way.

THE CHILD

A Taurean child can be surprisingly demanding, with a loud voice and stubborn nature. Plump, sturdy and strong, some are shy and retiring, while others may be bullies. This child can become completely absorbed in creative hobbies.

Artistic, musical and not particularly academic, Taurean children may prefer to sit and dream rather than struggle with mathematics. They must be encouraged to gain the basic academic qualifications that will enable them to pursue, later on, the creative courses they will enjoy. Practical and sensible, Taurean children often have a natural affinity for gardening, craft, decoration and any number of creative jobs.

FRIENDS

Sociability is your middle name and you chat to all and sundry at any social occasion. You make the person you are talking to feel they are the most important in the room for that moment. Your friends are usually the best type – helpful, loyal and happy to share the full social life that you enjoy. You may choose friends who have the same interests in music, fashion and the arts. You love a good gossip, so you can chat on the phone for ages. You are a great listener, so you help friends through their troubles and give sensible advice. However, you are not inclined to lend them money.

Foods
Any home-cooked food, peaches, fruit puddings

Indulgences
Cakes, sweet snacks

Exercise
Gardening, dancing, walking

Colours
Emerald or pink

Interests
Meals, holidays

Family activities
Cooking together

Flowers
Rose, pussy willow

Holiday
Sea cruise

LOVE

EMOTIONAL PATTERNS

You need a partner who boosts your confidence and offers security. As a child, you may have been made to feel slow, leaving you with a slight inferiority complex. You need a mate who appreciates you and makes you feel that you are good enough. Shared values are more important to you than sexual high jinks.

MARRIAGE AND PARTNERSHIPS

You are happiest when settled and rarely break up a marriage. If your partner wants to end your relationship, you resist with all your might. If the break-up occurs, you are likely to marry again. You tend to make your love relationship the centre of your life, possibly to the point of choosing not to have children.

LOVE AFFAIRS

These are not your style. While young, you may experiment, but your greatest need is a permanent relationship. If a marriage breaks down, you may play the field, but you really want to entice your partner back, or find a solid replacement.

COMPATIBILITY

Same-sign connections work well; many Taureans marry and live happily ever after. As regards another earth sign, a Virgoan could work, if they are not too restless; your shared interest in sensual sex would help. Capricorn could be a match, but might need too much attention.

You get on well with air signs, especially Libra; you could create a lovely home together. Gemini's practical, money-minded approach could work well, while the Aquarian irresponsibility with money would be intolerable.

Of the fire signs – Leo, Aries and Sagittarius – Leo might work, as you are both loving, loyal and attached to family. Aries are too interested in themselves to help you, while the Sagittarian need for freedom clashes with yours for security.

Water signs are a good choice, especially Cancer, who shares your interest in tradition and creative pursuits; Cancer's impulsiveness balances your preference for doing things thoroughly. Scorpio can be friendly and loving, if you both avoid obstinacy; you can allow their moods to pass without rising to the bait. Although Piscean impracticality may irritate you, shared interests in music, art and home-making give you plenty in common.

You are lucky – there are so many signs with whom you could be happy.

WINNING COMBINATIONS

Taurus with Scorpio and Cancer

WORK

WORKING ENVIRONMENT

There are areas of work that particularly suit your sign – working as a make-up artist or as a builder, for example. While these appear to be completely different from each other, there is an underlying similarity: the Taurean theme of creation and artistry runs through both careers.

Alternatively, your safe hands and love of beauty could take you into a job such as being a dancer or even working in the fashion industry. You also have a natural affinity with the production and presentation of food, so anything from farming to running a chocolate shop could fulfil you.

You may be attracted to the world of money and banking, but if you use your talents in this way, you will undoubtedly want to have a hands-on hobby for your spare time.

You have great visual ability, which makes you perfect for work in any kind of design capacity, whether this be engineering, bike mechanics, interior decorating or even landscape gardening. Beauty and symmetry in all its forms appeal to your sensitive sense of space and design.

YOUR ROLE AT WORK

You enjoy being part of a team in a fairly ordinary kind of organization. There is a side to you that yearns for glamour, so attaching yourself to the world of film or television in a technical capacity is another excellent idea.

JOBS

Chef, designer, builder.

SPENDING MONEY ON...

Home, garden, singing lessons

SAVING FOR...

A house with a garden

GEMINI

YOUR SUN SIGN

Gemini is a masculine air sign for which the symbol is the twins. You can do six things at once while chatting on the phone! In fact, you need a great deal of variety or you become restless. Many Geminis have two or more part-time jobs, and this probably suits your nature better than spending all your time in one.

Quick-witted and intelligent, you do things at a speed that others can only envy. Geminis have a good grasp of financial matters, and many work as accountants. Your greatest asset is the collection of skills and knowledge that you have stashed away in that amazing brain of yours. You use these to excel at quizzes or crossword puzzles. You don't lack intuition. Indeed, your ability to sum up people and situations at speed, coupled with an instinctive understanding of what is going on, makes you an excellent psychologist. The one area where this falls down is in your own life: while you can solve everybody else's problems, you find it difficult to get to grips with your own. You can be a terrible worrier and catastrophizer.

You have a reputation for flirtatiousness, but this is somewhat unfounded, since what others take for sexual interest is often no more than genuine curiosity and real friendliness. You love to chat and you enjoy parties that give you the opportunity to mix with a variety of people. Communicating with others is a major part of your working day as well as your social life.

You love to have lots of people dropping into your home, which means you are rarely alone for long. You may occasionally wish for peace and quiet, but the truth is that there is nothing you like better than a good laugh and a gossip with your friends.

Mercury, named after the Roman messenger of the gods, represents your love of speed and communication.

RULED BY MERCURY	YOUR ELEMENT
May 21 – June 20	Air

FAMILY

THE FATHER

A fairly easy-going man, the Gemini father copes well with fatherhood, but can become bored with home life. Some are so absorbed with work that they hardly see their offspring. At home, the Gemini father will provide books, educational toys and computers.

THE MOTHER

The Gemini woman can be pushy, as she sees education as the road to success. She encourages a child to pursue any interest and will sacrifice time and money for this. She is very busy at work.

THE CHILD

A Gemini child needs a lot of reassurance because they often feel like they don't fit in. They either do very well at school and incur the wrath of others, or they fail dismally but make up for this later in life. Some have excellent mechanical ability, others excel at sport.

Gemini children are quick and clever and such skills as mathematics and reading come early to them. However, many have such a hard time either at home or at school that they switch off and lose this head start. Nevertheless, there is a good chance these children will find their feet (and their confidence) and go on to further studies. Anything that enables a Gemini to communicate is a must, so excellent reading, writing and computer skills will be of great value. You might consider taking a journalism or marketing course when the time comes.

FRIENDS

Yours is a friendly sign. Getting out and about and sharing activities with friends is an important part of your life. If you are a keen sportsperson, you make friends among those who share your interests. You don't have a tremendous amount of confidence in yourself, so a good friend with a positive mental attitude can help your self-esteem.

You can be somewhat self-absorbed so you must remember to listen to your friends, especially when they have problems. Help them to deal with their troubles, just as you accept their support when you have worries of your own.

Foods
Pasta, seafood, light meals

Indulgences
Vaping, alcohol

Exercise
Gardening, walking the dog

Colours
Yellow or checks

Interests
Gossiping on the phone and on social media apps

Family activities
Gentle sports and ice- or roller-skating

Flowers
Rose, poppy, violet, foxglove, vine and clover

Holiday
Beach resort

LOVE

EMOTIONAL PATTERNS

Loneliness at home or at school may have plagued you during childhood. This could plunge you into a relationship much too early, when you haven't discovered who you are or what you need. As you develop emotional maturity, you'll look for someone who understands give-and-take.

MARRIAGE AND PARTNERSHIPS

Many Geminis jump into an early marriage. Unfortunately, such unions are often a repeat performance (or worse) of the dysfunctions of a typical Gemini childhood. As soon as you can, you leave this battleground and experiment a little. Once you finally settle into the right relationship, you feel safe for the first time in your life and will do anything to make your partner happy.

LOVE AFFAIRS

If you are lucky, you can find the right partner during your teens and then settle into a happy marriage. If your partner is selfish or even boring, you soon become restless, and before long set up one or more lovers outside the marriage.

COMPATIBILITY

You need a lot of emotional support, hard to come by from another Gemini. Either of the other air signs, Libra or Aquarius, make a better match, though the Libran must be laid-back and the Aquarian not too detached.

You could make a successful connection with a fire sign as they make you laugh. Aries and Leo can go out and earn a good living, releasing you from your fear of being without money. Sagittarius might be too unsettled for you, but you would enjoy travelling together.

Either Taurus or Capricorn, both earth signs, are reliable, practical and share your need for status: a large home with an income to match. The other earth sign, Virgo, has a similarly flexible attitude to life, but you are both born worriers.

Water signs are difficult for you. Cancer might just be feasible as this sign loves family life, but their moodiness would get you down. Pisceans are happiest living a chaotic lifestyle opposite to your methodical approach. You could be friends with a Scorpio as you share a sense of humour, but they would lose patience with you.

WINNING COMBINATIONS

Gemini with Aries and Libra

WORK

WORKING ENVIRONMENT

Variety is key for you, as you cannot stand working at a repetitive job. You can put your mind to a numbingly boring task if there is a specific reason for it, but once it is finished you won't repeat it! Office life suits your character well, preferably a busy office where phones are ringing and things are happening all around you. If you do spend any part of your working day alone, it is while you are driving from one appointment to another, and even here your phone is constantly pinging.

You are able to do a dozen things at once, all of them quickly and efficiently, and you seem to be able to cope with anything that is thrown at you, as long as it doesn't become too stressful. Your nerves are not strong, so the pressure of top management is probably not for you, but you really thrive as a team member in a busy, happy place.

Many Geminis travel around during the course of their work, which suits them perfectly. Although they are naturally very sociable and love working with people, they also love the sense of freedom that this brings them.

YOUR ROLE AT WORK
You do not like to be ordered around and you can cope with surprisingly large projects, so a management position suits you. Some Geminis are not ambitious but still like to feel needed, in which case a key role could be a good choice for you.

JOBS
Journalist, accountant

SPENDING MONEY ON...
Clothes, gadgets

SAVING FOR...
Your dream car

CANCER

YOUR SUN SIGN

Cancer is a feminine water sign for which the symbol is the crab. This makes you sensitive, sympathetic and caring. Interested in people, you are a good listener and would make an excellent counsellor.

Although you may look soft on the outside, you have a stubborn streak, which comes to the fore if anyone criticizes or attacks one of your family. You will bend over backwards to help a loved one who is in trouble.

Many Cancerians are shrewd business people, often with small businesses, especially shops, or agencies that help the public in some way. You will probably make a success of any kind of trading. It is typical of a Cancerian to be able to cope with chaos at work but to need peace at home. Yours is a very domestic sign and you may be a keen cook. You almost certainly like good food and wine, but are unlikely to be a glutton or a heavy drinker.

You can be clannish and suspicious of outsiders, preferring to keep both business and personal matters within the family. While you hate to waste money and may be penny-pinching or selfish in small ways, you are generous in big ones. Your relatives can count on your support when the going gets rough. You are probably an animal lover.

You tend to look backwards rather than forwards. Some of this is sentimentality, but you can also dwell too much on past hurts. History fascinates you and you may collect objects with unusual backgrounds. You don't throw much away.

You can be dreamy and your strong imagination can take you anywhere you want to go. Many Cancerians escape humdrum daily life by reading and dreaming of faraway places or

RULED BY THE MOON	**YOUR ELEMENT**
June 21 – July 22	Water

glittering business schemes. However, your cautious nature means you are unlikely to turn these dreams into reality, unless you have an adventurous partner who encourages you to do so.

Blues and greens represent the water associated with your sign. You have a longing for the sea and are drawn to it.

FAMILY

THE FATHER

A true family man, the Cancerian father embraces all his children. Cancerian sulks and bouts of childishness can be confusing or frightening for some children however, while his changeable attitude to money may make them unsure what they can ask for.

THE MOTHER

An excellent home-maker, the Cancerian mother is cheerful and reasonable, as long as she has an interest outside the house. She instinctively knows when a child is unhappy and can deal with that in an efficient, loving manner.

THE CHILD

Shy, cautious and slow to grow up, this child may achieve little at school and be swamped by louder, more demanding classmates. The Cancerian child may complain loudly about every ache and pain, or suffer from phobias.

Although Cancerian children are bright, they may not push themselves at school. The best you can do is encourage them to learn the basics and then develop their creative, artistic and musical talents. Anything slightly business-orientated is useful, because their canny nature will take them into business later in life. Those who want to look after children or animals should take courses that qualify them for these jobs.

FRIENDS

Yours is such a family-orientated sign that real friendships are few and far between. But you do need a friend outside, to allow yourself to let off a bit of steam and have a break from family pressures. Your friends must have a sense of humour and the ironic way of looking at life that you appreciate. If they enjoy singing, dancing and eating out, so much the better.

Your sense of humour means you can cheer anyone up, and your ability to listen makes you a wonderful friend. You enjoy trips with your pals, and a nice meal out and a chance to have a laugh do you the world of good.

Foods
Risotto, poultry, pudding with custard

Indulgences
Chocolate

Exercise
Swimming, horse riding

Colours
Silver or white

Interests
Eating out at good restaurants

Family activities
Family get-togethers

Flowers
Rhododendron and cactus

Holiday
The latest theme park

LOVE

EMOTIONAL PATTERNS

You carry unfinished childhood business into adult life. You may have grown up very attached to the parent of the opposite sex. You are not really interested in bed-hopping, but unless you strike lucky early, you can go through an amazing number of lovers in your search for the perfect parental replacement.

MARRIAGE AND PARTNERSHIPS

Yours is a cardinal sign, so you are not as easy-going as you look and need a partner who can cope with your demands. If you try to control a lover, you'll soon be on your own again. If you put them first and keep your sulking for really big issues, you'll have one of the best marriages around.

LOVE AFFAIRS

Cancerian love affairs are not carried out for fun, but to keep loneliness at bay. You are a pushover for someone who makes you laugh.

COMPATIBILITY

A relationship with another Cancerian works if you take turns with your moods and manipulative behaviour. Neither of you would come right out and ask for anything you need – but intuition might come to your rescue. Of the other water signs, Pisceans need lots of attention; fortunately, you love to do this. Scorpio likes to behave deviously, but you will see through it.

You are drawn to fire signs – it is an attraction of opposites. There are an amazing number of successful Cancer-Aries marriages, particularly when the Cancerian is female, as childlike Aries and motherly Cancer fit well. Other fire signs make great friends, but Leo won't be told what to do and Sagittarius can't be smothered, so marriage isn't an option.

The air signs – Gemini, Libra and Aquarius – don't understand you. They operate from a level of logic at odds with your intuitive nature. On the other hand, Libra loves to settle down and make a nice home. Of the earth signs, Capricorn and Taurus fit well and Capricorn shares your interest in business. Virgo, however, can't cope with your emotions. These three rely on you to make them feel safe and loved.

WINNING COMBINATIONS

Cancer with Taurus and Pisces

WORK

WORKING ENVIRONMENT

Chaos at work doesn't bother you as long as peace reigns in your home: you enjoy being in a busy environment with people around and plenty going on. Loud noises upset you though, so working in a noisy place such as a building site or airport might be a little too much for you. However, if you were to decide to work at a kennel, it wouldn't matter how many dogs were barking, because you'd love them all and you'd be able to distinguish between happy barks and distressed ones.

As well as loving animals, Cancerians are naturally friendly and genuinely interested in meeting a variety of new people. Any job which allows you either to travel around or to work face-to-face with the public would suit you very well – such as shop work or acting as a sales representative. You have a phenomenal memory, and faces and names rarely escape you. With your excellent memory and love of history, museum work would appeal to you. You like being on or by the sea, which can lead you to work for a travel organization or even to join the Navy.

Many Cancerians are happiest when running a small business of their own, so keeping a shop suits you well. Some Cancerians do extremely well by owning pet or antique shops, where they can offer their specialized knowledge to the public. Others are happy when working as estate agents or managers, due to the natural affinity between this sign and the property world.

YOUR ROLE AT WORK

Because you need your own space, running a small shop would suit you nicely. You can find it difficult to get on with colleagues at work, so a job that allows you to be independent may be a better idea. As you are very persuasive and you enjoy sales work, working in PR might be an appealing job.

JOBS

Pension planner, childminder, shop owner

SPENDING MONEY ON...

Family, travel, entertainment

SAVING FOR...

A sea-worthy boat

LEO

YOUR SUN SIGN

Leo is a masculine fire sign for which the symbol is the lion. You have high standards, not liking to let yourself or anyone else down. However, by trying to do everything to your best, you can be a perfectionist workaholic.

You can be ambitious, and when you have an idea there is little that will stop you. Although you can suddenly lose confidence in yourself when things go wrong, you usually bounce back. You are a high earner because you are a high spender and you cannot stand menial jobs, so you might as well become the boss. You don't aim to impress people and, as a result, you can't understand why others are sometimes jealous of your success. When times are rough it is your sense of humour that brings you back to normal. Although you can be irritable when things go wrong and sarcastic when under pressure or overtired, you rarely set out to hurt anyone. You certainly won't tolerate being treated with contempt or ridicule.

Leos are known for their generosity. If someone is in trouble, you will do all you can to help. You like 'the best' and prefer to go without rather than buy poor-quality items. When you are in the money, you can be extravagant, and there's nothing you like better than to eat out in a good restaurant with close friends.

There is a childlike quality about you and, while you are extremely businesslike outside the home, you love to curl up with your partner and be intimate in a joyful manner. You enjoy playing with your children, finding it easy to enter their world and be young with them. If you participate in sport, you enjoy success, but don't seek to win at all costs; having fun playing the game is more important to you than winning.

The brave lion symbolizes your big-hearted sign, and your ruler, the Sun, is king of the solar system.

RULED BY THE SUN	YOUR ELEMENT
July 23 – August 22	Fire

FAMILY

THE FATHER

Leo is a wonderful father as long as he remembers that children are not simply small adults. He likes to be involved with every aspect of his children's lives, encourages them to do well at school and happily makes sacrifices for them.

THE MOTHER

Though caring and responsible, the Leo mother is not satisfied with pure domesticity. She will put up with disruption, but becomes irritable when tired and may expect too much of her children.

THE CHILD

Leo children know from day one that they are special, loved and wanted, but also that their parents have high expectations, which puts pressure on them to be successful. They appear outgoing, but are sensitive and easily hurt.

Leo children tend to be very bright but incredibly lazy! They can be a perpetual source of frustration to their parents because they can do anything they set their minds to... but don't set their minds to anything. Don't despair – Leo children invariably educate themselves when they finally see a reason for it. Leos have an affinity with tech and, whatever field they wish to work in, soon pick up the necessary skills. Leos are suited to positions of leadership, so anything that trains them to manage staff is useful.

FRIENDS

Yours is a naturally friendly sign, and friendships form an important part of your life. You are probably the most loyal and helpful of all the zodiac signs and will pull out all the stops for a friend in trouble. Unfortunately, your generosity and loyalty can be misguided and you don't always get the support you want. Your friends are often weaker than you, which can make you feel powerful and needed.

You keep good friends for years and rarely let them down. If your pals find themselves in difficulties, you are the first person they call upon to give help or advice. You enjoy a laugh with your friends, especially if you play sport with them.

Foods
Smoked salmon, exotic fruits, Chinese food

Indulgences
Fancy cheese, chocolate éclairs

Exercise
Swimming, dancing, skating

Colours
Gold or orange

Interests
Shopping for clothes

Family activities
Visiting places of interest

Flowers
Lily of the valley, lavender, orchid and gladiolus

Holiday
Smart hotel in a lively resort

LOVE

EMOTIONAL PATTERNS

Some Leos experiment with partners before marriage, while others don't bother until ready to settle down. The chances are that you will have at least one serious relationship during your youth, with someone who lives far away or is not free. You choose emotionally needy partners, but also need to be able to admire them.

MARRIAGE AND PARTNERSHIPS

Because you are both independent and needy, you take a bit of understanding. You may be a huge success, but you're insecure and need a partner to help you achieve self-worth. If you find someone who understands this and loves you for it, you will blossom into the most happy spouse. If your partner resents your capability, you become miserable and walk away.

LOVE AFFAIRS

You need excitement, and unless there is a certain amount of stress, you feel empty and bored. Most Leos find this stimulation in their careers, but if all else fails – or if your marriage breaks up – love affairs will fill a short-term gap.

COMPATIBILITY

You find all the fire signs – Aries, Leo and Sagittarius – great as friends but of these, fellow Leos are the easiest to live with. Arians can be too inclined to want their own way, though Sagittarians are less confrontational.

Earth signs could make a good match, though Capricornians could be too involved in their careers to be real companions. Taurus and Virgo are both humorous and sensible, but you could lose patience with a neurotic Virgoan. You'll find the stability you crave with a Taurean.

Air signs are good partners, especially Libra who shares your desire for a comfortable lifestyle. Air signs share your strong sex drive and stop you getting bored, but you could find Gemini too self-obsessed. All air signs need generous partners, so make sure you are in a high income bracket!

The only water sign that works for you is Scorpio as, like you, it belongs to the fixed quality, so you can find the constancy you need. Neither Cancerians nor Pisceans think or act the way you do. Cancerians may be too emotional, but you do share a love for family life and may have mutual business interests.

WINNING COMBINATIONS

Leo with Aquarius and Scorpio

WORK

WORKING ENVIRONMENT

Despite the fact that some astrologists will tell you that you need a glamorous environment, you can do a dirty and menial job as long as the pay is good and – better still – as long as the business belongs to you! However, the fact is that you do prefer a prestigious role, such as working at a company that is leading in its field. You cope very well with stressful situations and impossible deadlines.

Your ultimate job would be that of a glamorous entertainer – a singer or movie star in the glittering world of show business. You love to befriend those who are powerful and influential, because the chances are that you are also in a position of power and prestige.

If you have your own business, you work extremely hard to make a success of it because you can't bear the thought of failure. Elegant clothes, a chauffeured limousine and business trips to glittering locations such as Tokyo and Las Vegas suit you very nicely.

YOUR ROLE AT WORK

There is only one position in any organization for you – being the boss. If you can't work your way to the top in someone else's organization, you simply consider the possibility of starting your own.

JOBS

Celebrity, high-flying executive, diamond merchant

SPENDING MONEY ON...

Car, clothes, perfume, jewellery

SAVING FOR...

A month at a spa

VIRGO

YOUR SUN SIGN

Virgo is a feminine earth sign for which the symbol is the virgin. Virgoans are often defined as cool, controlled, even sexless, but how can this be when your earth sign is associated with the harvest?

You have high standards of decency and self-discipline, even if something is boring. You are able to cope with painstaking work as long as you can stretch their mind elsewhere.

Shyness can make you appear standoffish, but once people know you they discover your wonderful sense of humour and kind heart. Cruelty disturbs you and you try hard to be kind to others; your super-sensitive nature makes you nervous around unkind people. If someone upsets you, your excellent grasp of language and quick mind mean you can be extremely hurtful.

Your lover will need to understand that you are not overly tactile. You prefer to show your love either in practical ways, by doing things for your loved one. You are happier in the background of life – in fact, your primary motivation is to serve – so you don't mind if your partner is more outgoing, or even a little bossy. You are an extremely loyal friend and like to keep friendships going for years. You particularly love encouraging others to try out those adventurous things that make you nervous, so you can vicariously enjoy their success.

Sport is not at the top of your list of activities – you'd rather read a good book. Most Virgoans are health-conscious, and you may even be a hypochondriac or a fussy eater. Some, however, seem to be the opposite; one of those polar situations that arise in astrology, where you belong to either one extreme or the other.

The virgin represents your gentle, careful nature. Yours is a hard-working and clean-living earth sign.

RULED BY MERCURY **August 23 – September 22**	**YOUR ELEMENT** **Earth**

FAMILY

THE FATHER

Embarrassed by open declarations of affection, the Virgoan father finds it hard to give hugs and reassurance to small children. Yet he loves his offspring dearly and will go to any lengths to see they have the best possible education and outdoor activities.

THE MOTHER

The Virgoan woman tries hard to be a good mother and loves her children very much. She may be fussy about details, such as dirt on the floor or the state of books. If she keeps a sense of proportion, she can be the most kindly and loving parent.

THE CHILD

Practical and capable, the Virgoan child can do well at school but may not always be happy. Sometimes they do not fit in and they may be slow to make friends. Shy and sensitive, they can find it hard to live up to their own high standards. However, their inquiring minds make them excellent pupils.

Virgo children are very bright, with excellent analytical skills. Reading and writing skills come early to them, and they usually end up with a good basic education. But nerves and a lack of confidence can hold these gentle children back, and they can be bullied, which may slow their progress down when they should be forging ahead. Gentle treatment and an encouragement of artistic or other creative activities will help build confidence.

FRIENDS

You are one of nature's communicators and need friends you can share ideas with. Your door is always open, and many friends pass through it, enjoying your generosity and company. You need this, because feelings of loneliness can strike you more quickly than they do others. You instinctively know that a good laugh with a trusted friend restores you.

You may think that you can't make a success of relationships, but your ability to make and keep friends proves you wrong. If you can only bring the same relaxed air to love relationships that you do to friendships, you will learn to heal past hurts and be truly happy.

Foods
Pizza, stir-fries or any quick and easy meals

Indulgences
Cookies, small cakes

Exercise
Badminton, gardening

Colours
Autumn colours

Interests
Books, art galleries

Family activities
Listening to music

Flowers
Pinks, rushes and pimpernels

Holiday
Tour of a historically troubled land

LOVE

EMOTIONAL PATTERNS

When you were a child, you may have felt that you didn't count for much or that you didn't fit in. As a result you have probably trained yourself to please others, or stay in the background. If you find a partner who makes you feel valued, then you blossom. You may find it difficult to be affectionate, so show your love by listening, talking, or sexually.

MARRIAGE AND PARTNERSHIPS

You can live without a permanent partner, as long as your life is filled with good friends. Sex is another matter, though, and you may have a host of short-term relationships or one-night-stands. Many Virgoans marry in their teens, happy to escape from family and make a life of their own.

LOVE AFFAIRS

You may never trust a partner enough to commit, preferring to keep your options open. Sex may be your substitute for affection. One thing that can kill any relationship is having to support a partner financially.

COMPATIBILITY

You can happily link up with your own sign, enjoying each other's quick minds and humour. You can be as interested in business as a Capricornian and as keen for security as a Taurean, but might find their inflexible attitudes hard to take.

The air signs – Gemini, Libra and Aquarius – could work well due to the logical way in which your minds work. Gemini's inclination to worry could be too close to yours for comfort. Librans can be unpredictable, and your sign is too sensitive to deal with this. Aquarius's incisive but intuitive intellect could complement your meticulous approach.

Fire signs frighten you! Arians can seem like brutes. Leos make good friends but, in a marriage, could make you feel shoved out of the way. Sagittarians deal with difficult times by finding humour in them, though you need more emotional support than they can offer.

Water signs – Pisces, Scorpio and Cancer – operate on a level of emotion that is hard for you to comprehend. However, there is often a reasonable connection with Pisces, if they are the fairly organized type. Scorpio could be a match, as you both love to make and save money. Cancer would be less positive, as you might not want to indulge in mother-and-child role-playing.

WINNING COMBINATIONS

Virgo with Sagittarius and Gemini

WORK

WORKING ENVIRONMENT

You may secretly feel that you are cleverer than those for whom you work and that you could do their job better than they do, but the fact is that you do not really feel comfortable in a position of power and responsibility. You need to be a step or two away from the top, so the real decision-making can be shouldered by someone whose nerves are steadier than yours.

You enjoy supplying a service or applying your brains for the benefit of others, but you need proper recognition and also generous payment for your efforts. Many Virgoans are freelance workers, either taking their skills with them from place to place or working from home and therefore meeting the needs of a number of large organizations. Freelance journalism or creating and selling software may suit you well.

Your sign is particularly keen on all aspects of healthy living so you are naturally drawn to the health industry, working in establishments such as health food shops and fitness centres. You love to encourage others to live a healthy life and to understand the need to eat properly. You may take up nursing and remain in the traditional area of the medical profession, or decide later on to work in a more alternative therapy field.

YOUR ROLE AT WORK

You are a wonderful assistant, but are uncomfortable in the role of boss. Giving orders to others doesn't come easily to you and – if they are slower and less efficient in what they do than you are – this irritates you.

JOBS

Writer, software programmer, nurse

SPENDING MONEY ON...

Books, music & films, medicine

SAVING FOR...

A year's sabbatical

LIBRA

YOUR SUN SIGN

Yours is a masculine air sign and your symbol is the scales. The scales suggest two different kinds of personality, both typically Libran. Venus, your planetary influence, is a beauty-loving, self-indulgent planet, while your sign is also cardinal, pointing to a go-getter. No wonder yours is such a confusing sign – and sometimes you are the most confused of anybody! The most obvious manifestation of this is your hatred of making decisions, or of being rushed into anything.

Some Librans want little more than a happy partnership, pleasant job and enough money to live on. All Librans can be laid back or even lazy at times (though what appears to be laziness is sometimes just indecisiveness). However, there are plenty whose energies come from the ambitious side and do very well.

An excellent conversationalist, you are welcome at any party because of your happy smile, friendly manner and general niceness. At home, though, you can be extremely grumpy, especially if you've had to hold back your feelings during the working day. Sometimes you can be downright unreasonable, and argumentative. Your talent for argument can be put to good use in negotiations, as you often spot good solutions for all parties involved. With your strong sense of justice, you can see both sides of an argument.

In fact, your friendly nature and humorous, non-hostile approach put others immediately at ease. This means you can assist people to deal with the trickiest situations and, at the same time, help to keep the emotional temperature around you down to a reasonable level. This makes you a wonderful advocate, diplomat or trouble-shooter.

Librans of both sexes are domesticated and happily run a household, even down to quite extensive DIY. You need nice

RULED BY VENUS
September 23 – October 22

YOUR ELEMENT
Air

surroundings and may spend a lot on gadgets, though television interests you less than music. You can be a wonderfully relaxed companion to your partner.

The only inanimate symbol of the zodiac, the Libran scales weigh up and consider both sides of the coin.

FAMILY

THE FATHER

Although a Libran father means well, he may slide out of the more irksome tasks by having an absorbing job or a series of hobbies to keep him occupied outside the home. He comes into his own with older children that he can talk to.

]THE MOTHER

Though pleasant and easy-going, the Libran mother is sometimes more interested in her looks, home furnishings and friends than her children. Others are loving and kind but a bit soft, which results in their children not respecting them and treating them badly in later life.

THE CHILD

Charming and attractive, the Libran child has no difficulty in getting on with people. They make just enough effort to get through school and to do chores they cannot avoid.

These children can be lazy and get by on looks and charm. Libran children must be pushed to gain the knowledge and qualifications they need. Often some time away from home in their late teens or early twenties can jolt a Libran into action and encourage them to stand on their own two feet.

FRIENDS

Genial, friendly and sociable, you make friends easily and happily get on with most people you encounter. Many of these friendships are superficial, but you do need at least one loyal long-term friend who goes back to your youth, who knows you better than anyone else, has a non-judgemental attitude and is sympathetic.

You enjoy friendships very much and they last for years. You may not make too much effort to keep in touch, but you do ring your pals occasionally and this prompts you all to get together from time to time. Meeting regularly once a week is a good idea.

Foods
Indian or Mexican foods

Indulgences
Alcohol

Exercise
Walking, tennis, golf

Colours
Pink or green

Interests
Comparing status symbols

Family activities
Long country walks

Flowers
Acanthus, lotus and wildflowers

Holiday
Villa in Tuscany

LOVE

EMOTIONAL PATTERNS

Most astrologists stress the importance Librans attach to partnerships, but many are happy to live alone. Yours is a confusing sign: you want marriage but also fear entrapment. Some Librans try to have both solid marriages and outside liaisons at the same time.

MARRIAGE AND PARTNERSHIPS

You seek partnerships in business, love and leisure, but are also extremely independent. You may be so keen to be part of a couple that you rush into an unsuitable marriage, and you have to guard against bullying a weaker partner, or wearing your lover down with trivial arguments.

LOVE AFFAIRS

Although you are a terrible flirt, you are an adept lover. Going out for meals, discovering shared tastes in music and travelling keep you happy. In an affair, you love your partner deeply while being relieved that they will be going back to their own home.

COMPATIBILITY

Although you love being in a partnership, your independence can lead to a lack of ability to commit. You could make a good match with another Libran who shares your free spirit. As regards the other two air signs, if you are a super-logical Libran a relationship with an Aquarian might work, but a typical Gemini neurotic would get you down.

Water signs are not a great match. While you could share creative interests with Pisces, you would find some too woolly-minded to live with. You could have a good friendship with Scorpio or Cancer, but living with those emotional signs wouldn't work. The earth signs are quite a good choice. Taurus shares many of your interests and there would be an excellent sexual connection, though you both have a strange attitude to money. Virgo might work well because they share your love of abstract ideas. Capricorn is a cardinal sign like your own, so there might be a battle for supremacy.

The fire signs – Aries, Leo and Sagittarius – generate an electric spark. The best would be Aries, as you would both enjoy your many fights! Leos are too obstinate and fond of their own way, but Sagittarius could be flexible enough to cope with your mercurial moods.

WINNING COMBINATIONS

Libra with Gemini and Aries

WORK

WORKING ENVIRONMENT

You are the world's best liaison officer. Agency work that allows you to put the right person or service into the right place gives you great satisfaction, so a career in recruitment, or a start-up that supplies specialized goods, could be just right for you. You love to give good advice and your retentive mind means that you usually know just where to lay your hands on the appropriate piece of information when it is needed.

Legal work suits you as you can always see both sides of an argument and can be sure to find a logical and fair way of dealing with any kind of legal matter. Many Librans are competent cooks, electricians and craft workers. They use their talents to start climbing the ladder of success, but no longer need these skills as they move beyond the hands-on level.

Your pleasant and accessible manner makes it easy for you to deal with the public, and people instinctively feel at ease with you and trust you completely. You have a good brain and fine judgement, rarely entering into work that is speculative or risky. If you work in a sales environment, you ensure that you know all there is to know about your product before you embark on selling it to others, because you realize that it is your knowledge and experience that will guarantee success. You are extremely conscientious about offering a good after-sales service as well, so your customers know that they will be looked after and receive good value for money.

YOUR ROLE AT WORK

Any position that allows you to smooth things over between people is a natural one for you. You love to be the 'front' person, the maitre d' in a restaurant or the PR chief of a political organization, where you can charm everyone into doing things the best way.

JOBS

Lawyer, employment agent, decorator

SPENDING MONEY ON...

Home, hobbies, outings

SAVING FOR...

Household refurbishments

SCORPIO

YOUR SUN SIGN

Scorpio is a feminine water sign for which the symbol is the scorpion. Although sometimes described as secretive and suspicious, the reality is that Scorpios are hardworking and honest, and their greatest interest is the happiness of their family. Of course, if you want to be sex-mad and subject to fits of uncontrollable jealousy, that is your prerogative!

A passion for life dominates your personality. You have enormous energy, and do not believe in half measures. Loyalty is your middle name. You give everything to your employers, and as long as they are decent to you, can stay happily in the same job for many years. You may reach the top of your profession, but are likely to prefer a second-in-command position, as you are happiest being the power behind the throne. Although honourable, you can often be manipulative.

Many Scorpios enjoy a dramatic life, and nobody is better than you in a crisis – perhaps it's this that makes so many Scorpios excellent nurses and doctors. At some point in your life you learned not to be free with your trust. You don't let others into the secrets of your bank account, private life or real feelings. Those who love you are probably the only ones who know when you are unhappy. Some Scorpios prefer animals to people, and even people-loving Scorpios love animals.

You may appear tough, but you are amazingly sensitive and can be badly hurt. You never forget a person who hurt you, or the circumstances of it, but you can forget that others can be sensitive too. Holding back that sarcastic remark is often a more successful way of operating than going on the attack. Be sure to let your delightfully soft side show sometimes.

Like your symbol the scorpion, you can be formidable if angered. The Roman god Pluto rules your sign.

RULED BY PLUTO & MARS	**YOUR ELEMENT**
October 23 – November 21	**Water**

FAMILY

THE FATHER

There are no half measures for a Scorpio father: he can either be really awful or absolutely wonderful. A Scorpio dad will provide love and security because he sticks closely to his family and is unlikely to do a disappearing act. Difficult Scorpio fathers can be loud and tyrannical.

THE MOTHER

Either the Scorpio woman is a wonderful mother or she has very little maternal instinct, despite trying to do her best. She encourages her offspring educationally and in their hobbies but has no time for children who whine or who are miserable.

THE CHILD

The Scorpio child is competitive and self-centred, and, when in an awkward mood, unwilling to cooperate with siblings, teachers or anyone else. They can be extremely hard to understand, living in a world of their own and harbouring all kinds of angry feelings.

Competitive and hardworking, Scorpio children do well once they have found a subject that interests them. A teacher who shows interest in a Scorpio child will be rewarded with faithful devotion. Sport and any hearty outdoor outlet is an absolute necessity for these energetic, tense children.

FRIENDS

You may be completely wrapped up in family or work and have no time for friendships. Or you may have a wide circle of acquaintances whose company you enjoy on a superficial level. Your best long-term friend is unfailingly loyal and, most importantly, will keep your secrets and share their secrets with you. If you are single, you enjoy 'dating' with a favoured friend.

You keep your friends for as long as possible and will put yourself out to help them. You may be a little more inclined to give advice than to listen to what a friend has to say, but your pals do rely on your strength; they know they can turn to you in times of trouble.

Foods
Soup, chicken, steak

Indulgences
Sweets, chocolate cake

Exercise
Football and hockey

Colours
Cherry or plum

Interests
Sport, dancing

Family activities
Visiting fun fairs and theme parks

Flowers
Pansy and ivy

Holiday
A car trip across Europe

LOVE

EMOTIONAL PATTERNS

Unless very atypical, Scorpios want to be happily settled. You must feel you can trust your partner completely. You need to dominate, and as long as this manifests as protective rather than possessive, a partner might appreciate your 'take-charge' attitude. You need someone you can open up to, so your childhood feelings of isolation can be forgotten.

MARRIAGE AND PARTNERSHIPS

Some Scorpios try marriage once, but end up happily living an independent life, finding affection when they feel the urge. Most, however, make a firm commitment. Yours is an all-or-nothing sign: if you decide to be faithful, cheating doesn't cross your mind; if you decide to sleep with as many people as you can, nothing stops you.

LOVE AFFAIRS

Before you find a partner, you are likely to have at least one intense affair. Once you are settled, affairs are off the menu. It is a rare Scorpio who spends their life skipping from one lover to the next.

COMPATIBILITY

The water signs – Cancer, Pisces and your own – are easy for you to live with as their emotional approach to life is like your own. The air signs don't operate in the way that you do, but you might want someone different from yourself. Aquarius is a fixed sign like you, so you're both reliable, though Aquarian impracticality might get on your nerves. Gemini's light touch could help you through attacks of the blues, but Libra's apparent softness, allied to a desire to get their own way, could confuse you.

Earth signs shouldn't present any problem. Taureans share your fixed purpose and need for security. Capricornians are ambitious, but you don't mind as long as they reciprocate in areas that matter to you. You share a sense of humour with Virgo, but your tendency to sulk would cause them to shut off.

Although fire signs attract you, the only one that works is Leo. Both are obstinate, both need to feel secure, and neither will walk out unless a relationship becomes impossible. Marriage with Aries would turn into a power struggle. As for Sagittarius, it's hard to see what would attract you in the first place.

WINNING COMBINATIONS

Scorpio with Leo and Capricorn

WORK

WORKING ENVIRONMENT

You are a hard worker and you need what you do to be meaningful, so simply serving cups of coffee won't fit the bill. Being proud, you need a position that has some serious status attached to it, though you may not actually wish to be the head of an organization. Your favourite position is one where your influence is great but your face and name are less well known than the charismatic character whose leadership you foster. You can be extremely ruthless if someone gets in your way, so the construction and destruction of other colleagues are equally easy for you to handle.

Many Scorpios start out with medical training, and whether you end up directly involved in this field or not, the urge to help and to heal others remains with you throughout life. Some Scorpios also have an instinct for all things mechanical, which may lead naturally into a career in engineering.

YOUR ROLE AT WORK

Being in charge of anything worries you and you are far too tense to roll with the punches, but you make an admirable second in command. If you can work for or alongside someone you really trust, your intuition and common sense will add much to the success of the enterprise.

JOBS

Doctor, soldier, management consultant

SPENDING MONEY ON...

Travel, music equipment

SAVING FOR...

A holiday apartment

SAGITTARIUS

YOUR SUN SIGN

Yours is a masculine fire sign and your symbol is the centaur, also called the archer. Because you need freedom, you dislike anyone asking where you are going or how long you will be out. Nor can you cope with people who want to run your life.

A lot of Sagittarians have jobs that take them travelling. You enjoy the variety of new faces you encounter during a working day. If you work in one place, the job must provide plenty of challenges as well as the opportunity to talk to many different people. With your love of education, further education could be ideal. Friends and family are important to you and you love to have loved ones come stay with you.

With your progressive outlook, you don't dwell much on the past. Your fine mind ensures you keep learning and taking an interest in everything around you. Many Sagittarians have a deep interest in philosophy, often rejecting the religion they were brought up with to find something with meaning for them. Astrology and spirituality appeal to many and lots of Sagittarians are excellent clairvoyants or mediums.

You are happy-go-lucky with an excellent sense of humour. You are likely to be good at sport and games, and could find an hour or two at the gym a good outlet for your considerable energies. Under stress, you can be sarcastic and hurtful, but this is usually fully compensated for by your terrific sense of humour.

Many Sagittarians are highly skilled DIY workers, able to put together a house almost from the bottom up. As well as great dexterity, you probably have an artistic eye, so whatever you turn out looks good and stands the test of time.

Your quest for knowledge is represented by the mythological centaur, Cheiron, an expert on philosophy.

RULED BY JUPITER	**YOUR ELEMENT**
November 22 – December 21	Fire

FAMILY

THE FATHER

Though he may not be at home much and may be somewhat unavailable to his children at times, the Sagittarian father will give his children all the education they can take, providing books, equipment and interesting excursions.

THE MOTHER

The Sagittarian mother is kind and easygoing. She may be very conventional, or unbelievably eccentric, filling the house with weird and wonderful people.

THE CHILD

The Sagittarian child loves animals and outdoor life but also loves to sit around on screens. They have plenty of friends, whom they rush out to visit at every opportunity. Happy and optimistic but highly independent, they cannot be pushed in any direction.

Sagittarian children are quick, clever and studious as long as they have plenty of opportunity to let off steam by playing sport. Charming and outgoing, these children get on well with their classmates. They are rarely afraid of teachers or adults, so there is little to halt their progress. Sagittarian children love a variety of unusual subjects, so need to be encouraged to do the boring basics to have the skills to bring to their special interests. Many are quite attuned to electronics.

FRIENDS

You are everybody's pal... and sometimes also nobody's. You have so many friends that it is hard for you to keep up with them. Those who want your friendship to be exclusive, or stand the test of time, will be disappointed because you tend to move on. But for those who are happy to pick up the threads once you are back in town, friendships can be kept going for years.

Your great value as a friend is that you can talk about anything and listen to anything without being judgemental or looking down on anyone. You like people, and that comes across to them. If somebody needs cheering up, there's nobody more able to make them laugh than you.

Foods
Sushi, exotic fruits and vegetables

Indulgences
Wine

Exercise
Climbing, team games, horse riding

Colours
Royal blue or purple

Interests
Sport, making new friends, travelling

Family activities
Watching your favourite sports team live

Flowers
Marigold, nasturtium, sunflower and cyclamen

Holiday
Resort with plenty of sports and activities

LOVE

EMOTIONAL PATTERNS

You were a joy to bring up as your parents hardly saw you. As soon as you could toddle, you visited friends. Your own family were a known quantity; to a Sagittarian, sitting with a known quantity is like reading the same book over and again. New faces and places were, and still are, food and drink to you.

MARRIAGE AND PARTNERSHIPS

Many Sagittarians avoid marriage. Those who marry early sometimes walk away to live alone for a long spell; they may remarry later, when their restless nature calms down. For you, boredom finishes off any relationship.

LOVE AFFAIRS

The main attraction for you is a shared sense of humour. You relate easily to others and never struggle to find a lover. You can confuse others, however, because your greatest need is friendship. If a friendship also includes sex, your 'friend' may believe that the relationship means more to you than it really does.

COMPATIBILITY

You find it easy to get on with your own sign, as you give each other space. The other fire signs of Leo and Aries would also be good matches as you would fit in easily with their ambitions. Although Leo might not share your sense of humour, you both love the good things in life. Aries and you are both spontaneous, and you would stand up for yourself.

Air signs would make a match, as you could share intellectual ideas, although Librans can be confusing. Geminis are similar to you, except you love travel more than they. Aquarians have as many off-the-wall interests as you, and as long as these were similar, you'd get on well. Earth signs are too rooted for you, although Virgo could be a good intellectual match. Taurus wants to know that each day is going to be like another: diametrically opposed to what you want. Capricorn is ambitious, and you could happily spend their money!

Water signs puzzle you but there could be a connection with Pisceans, who need space as much as you do. However, you couldn't tolerate Cancer or Scorpio's attempts to dominate by manipulation.

WINNING COMBINATIONS

Sagittarius with Virgo and Gemini

WORK

WORKING ENVIRONMENT

Many Sagittarians work in the fields of broadcasting, publishing and the travel trade. It is the communication of ideas and the transport of people that attract you, because something in your psyche hates for ideas, people, situations or you yourself to stagnate for long in one place.

Your favourite environment is one that has many rooms, studios or offices, and lots of people popping in and out of them. You enjoy keeping up with the latest gossip and you always have your finger on the pulse of the organization you work for. Teaching is also another great job for you and the busy environment you find in a school suits you. An airport has a similar feel to it, being noisy and bustling, and the work is not too predictable. Silence and solitude would drive you crazy.

Some Sagittarians have excellent manual skills, which could open up the world of carpentry, electrical work or plumbing. If you do work in a technical environment, you do so extremely well, and get it right every time. You also have a highly developed aesthetic sense, that can lead you into the creative realm of art and design.

YOUR ROLE AT WORK

Although yours is a restless sign, you often find a role that suits you and then stick to it for years. You love to be part of a creative team where you can use your ideas to inspire others and you derive great satisfaction from seeing your plans put into action.

JOBS

Travel consultant, carpenter, broadcaster

SPENDING MONEY ON...

Sport, travel, your friends

SAVING FOR...

Three months trekking in the Himalayas

CAPRICORN

YOUR SUN SIGN

Capricorn is a feminine earth sign for which the symbol is the goat. Ambitious, prudent and hardworking, some can be too serious for their own good, though many can let their hair down. Some are painfully shy while others are outgoing, but even the shy ones become confident in middle age.

You don't enjoy being rushed into things. You are clever but use your intelligence in a practical way, and are not overly cerebral nor an intellectual snob. Your brightness, coupled with your capacity for hard work, can bring success in any number of fields. You can be tough in business but your polite manner and gentle humour help you get your way without upsetting others.

Your tendency to be too money-minded may alienate others or leave your family wondering when you will find time for them. You must try to create a balance in your life that allows for relaxation as well as work. This may be hard for you. Once you feel secure financially, your thoughts turn to exploration. You don't enjoy travelling alone, preferring to take your partner.

You take a responsible attitude to family life, and would do a good job of looking after step-children or in-laws. If you are a typical Capricorn, you are extremely good looking, with the fine bone structure ensuring you stay that way into old age. You can be a flirt, but if you are in a relationship you are intensely loyal.

Some astrologists write you off as a dour, stingy bore, but this is far from the truth. Although you hate to spend recklessly, the thought of a great party gets you loosening your purse strings faster than spendthrift Aries. You love music, are light on your feet, with natural rhythm – the side that loves to have fun!

The mountain goat is the symbol for your industrious sign, always striving to reach the top of the hill.

RULED BY SATURN December 22 – January 20	**YOUR ELEMENT** Earth

FAMILY

THE FATHER

A true family man who copes with housework, the Capricornian father is dutiful and caring, unlikely to run off, or leave his family wanting. He can, however, be too strict.

THE MOTHER

A Capricornian woman is a good mother but may be inclined to fuss. Being ambitious, she wants her children to do well and teaches them to respect teachers.

THE CHILD

A little adult right from birth, the Capricornian child doesn't need much discipline to do well at school. Modest and well-behaved, these children may be nervous and solitary.

Capricornian children are a teacher's dream as they are quiet, studious and bright. They listen well, do their homework and pass exams with ease. However, they lack confidence, and as they do not like contact sports, they take part in other activities such as gymnastics, dancing or horse riding as a confidence-building exercise. They must also be encouraged not to worry too much, or become upset if they do fail the occasional test.

FRIENDS

You may miss out on friendships as a result of being so wrapped up in your family or your career that you have little time to gossip or see pals. You take everything in life so seriously that even friendship is not something to be entered into lightly. Your friends also share your business interests.

Friendship for its own sake may not come easily to you, as it is always more natural for you to go out with work colleagues than to make time for friends who have nothing to do with your home life or your job. But if you make the effort, you will soon see just how rewarding friendship can be.

LOVE

EMOTIONAL PATTERNS

Many Capricornians have long relationships which come to nothing. Perhaps a lack of emotional confidence leads you to pick someone who can never really be there for you. After time spent alone, assessing your needs and developing your self-worth, you can find the loving partner you need.

MARRIAGE AND RELATIONSHIPS

Virtually the only thing that prevents you from forming a loving relationship is a demanding parent. You may become trapped by a deep sense of responsibility towards your family. Conversely, you must avoid spending so much time at work that you neglect your family.

Foods
Chicken, lamb chops, light puddings

Indulgences
Junk food

Exercise
Gardening, swimming, dancing

Colours
Black, brown or grey

Interests
Business interests

Family activities
Dancing and singing

Flowers
Orchids and buttercups

Holiday
Weekend city break

LOVE AFFAIRS

Love affairs are not your style, as you take love and sex seriously. If you do have an affair, your lover will probably be older than you, or of a high status.

COMPATIBILITY

You could find a successful relationship among earth signs, especially another Capricorn who would share your tact and encourage your work aspirations, though this could lead to boredom. Taurus would be a good match, as their artistic talents would lift your spirits. Virgoans are bright and humorous, but, like you, can be fussy.

A water sign partner would be better, as their emotional approach would balance your practicality. A Cancerian would share your interest in business and might have that touch of charisma that you lack. Scorpio could be an excellent choice, as you can tune out their demands while enjoying their warmth, sexuality and humour. Pisceans are very different, being artistic and perceptive, but this could be the thing that keeps your attraction fresh. Air signs are not a bad combination, though clashes would occur if you tried to make a life with a Libran. Gemini shares your tendency to worry so you would understand each other's feelings. You share with Aquarius an eye for detail; you would have to be the major money earner, though.

The impatience of fire signs would get you down. A Leo might be possible but an Arian would fight for their own way, while a Sagittarian wouldn't want to settle down.

WINNING COMBINATIONS

Capricorn with Scorpio and Taurus

WORK

WORKING ENVIRONMENT

Status and prestige are very important to you, and, while your own job may not be particularly exciting, you prefer it to be in an environment that deals with inspiring or interesting products. Many Capricornians work in publishing. Shipping is another area that might attract you, since it involves large quantities of goods being moved from one place to another. Similarly, banking or stock market analysis, where huge figures are 'crunched' through computer systems, can suit your outlook.

Dirty work in an outdoor environment is not for you and neither is any field that brings you up against harsh reality, such as spending your days inside an operating theatre. A nice quiet office where you could bury yourself in the accounts department would be much more appealing to you.

Many Capricornians have a unique nature that probably results from the fact that their sensible sign is wedged between the super-eccentric signs of Sagittarius and Aquarius. This can lead you into the fields of alternative health, with osteopathy heading the list of potential careers. These days there is no shortage of Capricornians working in aromatherapy and many other fields of healing and wellness.

YOUR ROLE AT WORK

Any job that involves figures, statistics and money works well for you. You may be drawn to a creative field such as publishing or theatre management, but you will still run this as though it were a banking organization.

JOBS

Banker, publisher, theatre owner

SPENDING MONEY ON...

Home, travel, investments

SAVING FOR...

Your pension plan

AQUARIUS

YOUR SUN SIGN

Yours is a masculine air sign for which the symbol is the water carrier. Aquarians are different from just about everyone else, including all other Aquarians, which obviously makes you hard to categorize.

You are extremely independent and your mode of thinking is totally individual. You value education highly, not only as the road to success, but also for its own sake. There is little that doesn't interest you and you love to be surrounded by friends chatting about a variety of subjects. If you have a family, you ensure they get all the education they can.

Aquarians can be found in any number of careers, but a job that involves communication is bound to be preferred. You are drawn to work in the public sector and in technical jobs, teaching or writing, and may prefer to be self-employed. As the sign of the zodiac with the most original mind, you may spend your life solving problems or inventing new methods.

Despite the intellectual reputation of your sign, you also have a strong physical side. Many Aquarians work as engineers or building craftsmen. Many members of your sign are very good at the individual sports – badminton, rally driving or tennis.

You have a broad mind and an easygoing attitude to other people's eccentricities. Although you do not necessarily see yourself as eccentric, you recognize that you are different from the herd. The sign of Aquarius is particularly associated with astrology, and other typical interests are the environment, alternative health and healing.

You look into everything and take every idea seriously, but then you form your own opinions, which are not easily changed once you settle on them. Like the other two air signs, you have

RULED BY URANUS & SATURN	**YOUR ELEMENT**
January 21 – February 19	**Air**

perfected the art of argument. However, you rarely do this to hurt others, only when you need to protect yourself, or want to have fun in debate.

The symbol for your air sign shows a vessel full of ideas and knowledge that you love to share with others.

FAMILY

THE FATHER

Though he may have no great desire to be a father, the Aquarian father makes a reasonable job of it. Some will even spend hours inventing games and toys, and all value education.

THE MOTHER

This mother may be too busy changing the world to see what is going on in her own family. Kind, reasonable and keen on education, she respects a child's dignity.

THE CHILD

Easily bored, an Aquarian child may be demanding when very small but much more reasonable once at school. They have many friends and may spend more time in other people's homes than in their own. They are stubborn and determined.

Extremely clever, Aquarian children love anything innovative. These children take to education like ducks to water, but their oddball attitude may make it hard for them to fit into a traditional school setting. Education doesn't stop at the school gates for them, so take them to museums or events, or give them books that reflect their special interests. History, geography and the liberal arts fascinate them.

FRIENDS

Your sign is synonymous with friendship and most Aquarians have many friends and acquaintances of all descriptions. Your friends are drawn from many walks of life, but the chances are that they are highly intelligent, have a good sense of humour and are not particularly demanding of your time or energies, though a friend who needs good advice can always come to you.

Your best friends must make the effort to keep in touch with you. You are one of the most friendly people in the zodiac, but you don't like your friends to feel as though they own you. You need people who will come and go in your life, enjoying a little of your company before you both drift off to pastures new.

Foods
The newest food trends, anything 'different'

Indulgences
Salty snacks, olives, pickles

Exercise
Golf, squash, cycling

Colours
Neon colours, electric blue

Interests
Committee work

Family activities
Fishing and skiing

Flowers
Cornflower, snowdrop, lily and narcissus

Holiday
Fishing trip

LOVE

EMOTIONAL PATTERNS

You are a sincere lover who means well, but find it hard to keep a relationship going. Yours is an awkward nature and you live in an unconventional fashion that partners find hard to cope with. You need group activities both in and out of the home. You love parties and like to chat and even flirt a little with friends or their partners.

MARRIAGE AND RELATIONSHIPS

Your sign is so freedom-orientated that one might expect you to avoid marriage, but this is rarely the case. Even if you make a mistake first time around, you'll try again until you find the right kind of oddball with whom to share your life.

LOVE AFFAIRS

You are not particularly keen on love affairs; you prefer either to live alone and have loads of good friends, or to be in a settled relationship. If you do have a short fling, you might be lonely.

COMPATIBILITY

Marriages between Aquarians work well, as you tend to have the same interests, along with your generosity and sense of humour. Either fellow air sign of Gemini and Libra could be a good match: Gemini would supply an interesting social life. Librans want their own way, which might lead you to behave obstinately.

Water signs attract you but also irritate you intensely, especially Pisces, whose mind doesn't work in any way you can relate to. Cancer wouldn't understand your need for independence, but Scorpio can work because your signs share intellectual curiosity.

Of the earth signs, Taureans need to keep everything under control – including you. A Virgoan's intellect could appeal, and their business-like attitude would complement your intuitive approach. Capricorn would be a good match as you could teach them to lighten up, while they would supply some practicality you lack.

Fire signs could work, especially Sagittarius, whose intellect and energy fascinate and inspire you. Leo's materialism might get you down. Your supportiveness may be just what Aries needs to counteract their occasional loss of confidence. An Arian, who may be impulsive, would defer to your logic and intelligence.

WINNING COMBINATIONS

Aquarius with Leo and Sagittarius

WORK

WORKING ENVIRONMENT

Although you are drawn to secure jobs in offices, you have a side to your character that needs to find a way to exercise your intelligence and originality.

Computing is second nature to you, and, while everyone else struggles to keep up with the latest technology, your analytical mind instinctively understands how it all works. Your grasp of all things technical and logical can lead you into computer graphics or the more technical aspects of astrology.

Teaching is a natural outlet for your talents, as it gives you a chance to help and inspire others and is also a good opportunity to show off your own extensive knowledge.

A great many self-employed builders, electricians and plumbers are born under your sign, and it is the feeling of running your own show and the freedom that this brings you that lure so many of you into these fields.

A safe, secure job making desserts in a large catering organization is not ideal for you.

YOUR ROLE AT WORK

In theory, you should be an independent worker, creating computer games or packaging designs but, in fact, this is rarely the case. Teaching is a good area for you to work in and you are most comfortable in a fairly senior management position.

JOBS

Inventor, astrologer, designer

SPENDING MONEY ON...

Books, computers, DIY tools

SAVING FOR...

A fully paid-up second home

PISCES

YOUR SUN SIGN

Pisces is a feminine water sign for which the symbol is two fish, tied together and swimming in opposite directions. Yours is a strange nature, as it carries within it so many contradictions. You can be efficient, capable and an excellent business person... but you can also be chaotic and muddled.

It is sometimes hard for others to see how you manage to switch between these two behaviour patterns. You are extremely sensitive and highly intuitive and there are times when this psychic ability takes over your more logical side.

Your nature is strongly sympathetic towards others, and you rush to help those in difficulties, sometimes at the expense of your own needs. Most astrologists don't mention the fact that you can be bossy! However, your motives are purely altruistic, as you simply love to encourage – or push – friends and relatives into taking steps that will make their lives more successful. You rarely think of your own needs, because you get so much out of helping others to reach their potential. You have more courage than expected, and when your intuition tells you something will work, you back yourself, coming out the winner.

Many Pisceans have an artistic streak, and can, as a result, not earn much. Your values are more spiritual than material, and the need to satisfy your creative urge more important than earning a high income. There is a strong mystical side to you that draws you to religious, spiritual or New Age ideas. A high proportion of Pisceans work in the wellness sphere, or even as mediums. Your planetary influence, Neptune, rules the feet, so one natural outlet for your talents would be reflexology, which involves treating the feet.

The colours of the sea reflect the unexplored depths that never fail to fascinate your contradictory sign.

RULED BY JUPITER & NEPTUNE
February 20 – March 19

YOUR ELEMENT
Water

FAMILY

THE FATHER

A Piscean father tends to fall into one of two categories. He may be kind and gentle and happy to take his children on outings and to introduce them to art, culture, music or sport; or he may be disorganized and unpredictable.

THE MOTHER

While she may be lax and absent-minded, the Piscean mother loves her children and is loved in return. Her children prosper, though many learn to reverse the mother-child roles.

THE CHILD

This sensitive child may find life difficult among more demanding brothers and sisters. They may drive their parents crazy with their dreamy attitude and they can make a fuss over nothing. They need a secure and loving home with parents who shield them from harsh reality.

These dreamy children can miss out at school because they sit looking out the window and tuning out. However, they can be excellent students if inspired by a good teacher who recognizes their talents. You can do much by encouraging a Pisces child to develop their considerable creative gifts. Piscean children need to develop organizational skills and must be firmly reminded to finish what they start, as they are prone to giving up if they lose interest.

FRIENDS

You attract two kinds of friend: the first needs your help and advice, the second is an oddball like yourself. People whose attitudes to life are staid or materialistic don't attract you, but those whose outlook is spiritual or humanitarian will find a place in your heart. Your friends should be the easy-going type who don't make a fuss in public when things go wrong.

Your friendships are deeply important to you, and if you don't have much in the way of a family life, friends can fill in the gaps. You are a thoughtful friend and always there when needed. However, you must ensure that you are not taken advantage of by those with stronger or more outgoing characters.

Foods
Fish, pies, any home cooking

Indulgences
Chips and crisps

Exercise
Dancing, swimming

Colours
Sea blues and greens

Interests
Evening classes

Family activities
Visiting places of historical interest

Flowers
Water lily and poppy

Holiday
Going on safari

LOVE

EMOTIONAL PATTERNS

You may only settle down once you have done all the experimenting you want. You love domestic life, but need some chaos to keep boredom at bay. Guard against letting a desire to be needed become bossiness, and resist the temptation to live other people's lives for them, while neglecting your own.

MARRIAGE AND RELATIONSHIPS

Pisceans generally turn to marriage early, and children come along quickly. Your tendency to view a partner though rose-coloured glasses at the start of a relationship can lead you to make the wrong choice the first time. You will form another relationship, or settle for a life filled with children, pets and hobbies.

LOVE AFFAIRS

You can go through an amazing number of affairs in your search for perfection. Along the way you may have lovers who activate your sympathy, but these one-sided affairs never last. You may settle for an imperfect partner – or keep on looking.

COMPATIBILITY

You need to be loved, supported and cared for. Two Pisceans can create a marriage either made in heaven or hell! If you are both stable, you can get along well, as Pisces is a sexy, humorous sign. Either of the other water signs of Cancer and Scorpio would understand you. Both can be moody... but so can you.

You won't find much support among fire signs, except perhaps Leo. You can form wonderful friendships with Sagittarians if you don't expect them to run around after you. Arians can be great colleagues, but their self-centred attitude would get you down.

Air signs' approach to life is so different from yours. Aquarians share your interests but a lifelong relationship with this cool, logical sign probably wouldn't work. Gemini is a mutable sign like yours, so can adapt to circumstances. Libra is a good choice as you share a strong imagination and love of creative pursuits.

Earth signs could provide stability without manipulating or smothering you. Taurean creativity would fit well with your ability to create a lovely home. Capricornians can be fun and are stable and reliable. Virgo would stimulate your mind; they might find you bossy, but probably won't mind too much.

WINNING COMBINATIONS

Pisces with Sagittarius and Scorpio

WORK

WORKING ENVIRONMENT

There are extremely efficient and capable Pisceans who make excellent businessmen and women, but as this is one of the most contradictory signs, there are also those who work in exactly the opposite way. There is a kind of shapeless, unformed aspect to the way these other Pisceans live and work, which must be poured into a creative outlet.

Your sign is full of artists, musicians, china restorers, sculptors, fashion designers, actors and dancers, all of whom use their creative talents to bring beauty into the lives of others. Your task is to remind us that the mundane day-to-day world we live in must be filled with light, colour, glamour and fun. You love the beauty and aesthetic appeal of crystals, and may work with these, if not in a healing capacity, then in the manufacture of jewellery or art. There are so many people who have the sun or moon in Pisces who work in the healing professions that it seems to be a prerequisite for the job.

YOUR ROLE AT WORK

You need to enjoy your working day and to feel that the products or the ideas you are handling mean something. Money alone is not a driving force in your life, but you do have a strong need to have your talents recognized and be appreciated by others. You could work alone or in a small team as an artist, jeweller or charity organizer. Whatever you choose to do, it should be creative and unusual.

JOBS

Artist, musician, wellness teacher

SPENDING MONEY ON...

Travel, *objets d'art*, self-improvement fairs

SAVING FOR...

A luxury recreational vehicle

DOMESTIC COMPATIBILITY

	ARIES	TAURUS	GEMINI	CANCER	LEO	VIRGO
ARIES WITH...	You'd better allocate responsibilities to each other right from the start, so each can have their own territory.	Taurus shouldn't expect Aries to be tidy in the home, but Aries will do the gardening.	Gemini needs a tidy home, Aries doesn't. Compromise: get some domestic help.	Fine, as long as Cancer is in charge of the home. Aries can do the gardening.	A fight for supremacy in the home! Aries argues and balks at giving up control.	Very difficult; Virgo is fixed and will drive Aries crazy! Aries needs variety.
TAURUS WITH...	Aries's untidiness will drive Taurus nuts, while Aries will wonder what it's all about!	A very comfortable arrangement for both. They will happily create a lovely environment.	Taurus takes care of Gemini, and Gemini appreciates and reciprocates that. Works well.	A similar outlook, as both are family people and will make the most of their home.	Leo will get bored unless Taurus makes an effort to keep a fresh, interesting environment.	Not bad, especially if Taurus does the cooking. The home will be neat and tidy.
GEMINI WITH...	Gemini must take charge, and Aries will happily let them do so.	It's OK, if Taurus keeps house and Gemini provides the fun. These two signs make a good pair.	Books, films, computers and gadgets keep both happy. Things will work comfortably.	If Cancer looks after the house (no problem!), Gemini will be happy. A light, bright atmosphere.	Both are relaxed home-makers, so no disagreements here. Some outside help would be appreciated.	Both value books and music more than expensive possessions, and will make a good home.
CANCER WITH...	Cancer will enjoy running the home, and Aries will be glad to go along with that.	Will usually work out well, with input from both. The family will have a stable environment.	This will be a neat and well-kept scene. Both like it that way, especially if Cancer is the motivator.	Like-minded, these two will be comfortable and practical. The children will have to behave!	Leo may be too lavish for Cancer, but the kids will love the home and the atmosphere.	No problems here, and the two will help each other out. Each will learn new tricks, too.
LEO WITH...	As long as Aries rules the roost at home, Leo can take the limelight in public.	Tensions predominate – there are basic differences on which neither wants to budge.	This can be a very satisfactory set-up. Both like an orderly home and youngsters will be well cared for.	There are lessons for each to learn and, for once, Leo will listen, as long as there isn't a critical attitude.	At home, mutual admiration prevails. In public, there's some drama, but it's usually over quickly. There's a strong chemistry between you.	This often works well with more mature couples. They know how to accentuate the positives.
VIRGO WITH...	Aries's relaxed attitude to household mess will drive Virgo to distraction. Virgo is too fussy for Aries.	Both signs love the comfort of home, but will they fight over who gets to choose the decor?	Both love lots of talk and a house full of visitors, but might fight over the cost of entertaining.	Both need a quiet and settled atmosphere in the home and both love family and friends around.	Leo will dictate terms and run the home, while Virgo will be stuck with all the DIY.	Both will work at making the household pleasant and both like to chat for hours.

LIBRA	SCORPIO	SAGITTARIUS	CAPRICORN	AQUARIUS	PISCES
Apart from fierce fights, not too bad. Both want their own way, but can find common ground.	One or the other: either total harmony, or a permanent battleground. Find out first, and fast!	Neither will be indoors very much – both like to go out. Definitely get some domestic help!	This could work, as neither is overly house-proud. Try to share the chores, to ease the burden.	Either great harmony, or open warfare. Aquarius is likely to be leaned on, but won't budge.	The relationship won't get as far as living together; they will both find that out.
A happy combination, as both are great homemakers. They will share the same ideas.	Scorpio will boss Taurus around, and Taurus will fight back. A difficult situation.	Totally different outlooks make this very difficult without some compatible planetary influences.	They both enjoy going over their bank balances. Both are practical, and should get on well.	Obstinacy on both sides stops this from working out. Both like their own way.	Well, well; a home filled with music and beauty. Great! There can be complementary fusion.
Gemini resists Libra's tendency to rule the roost. Perhaps let each rule part of the roost?	Scorpio will bully Gemini, until Gemini leaves home. It takes a strong person to live with that.	Neither can sit around indoors for long, and they will both be happy with that.	Both would enjoy running a business from home.	Both love company. The home would be a popular meeting place for interesting people.	Pisces's chaos in the home upsets and confuses Gemini. That only confuses Pisces further.
Conflicting ideas can be problematic. Libra wants things their way, and vice versa!	Cancer needs to compromise a bit here, but they'll understand each other's needs.	Whither away, Sagittarius? Dost thou leave thy partner alone at home again?	To learn from each other's opposite nature is the ideal. As spouses, they will do so.	Not a comfortable match. Cancer can't understand the erratic tastes, while Aquarius must have variety.	This is fine, Cancer will usually be happy to clear up the mess, as long as it's not unreasonable.
Don't upstage Leo – this is an unusually good relationship, and is worth working on!	Two fixed, too fixed. Scorpio will seethe and eventually sting the kingly paw.	Easygoing, fun, not difficult to allocate responsibilities. A lively, attractive home.	The home will tend to take on Leo's style. There will be distinct opinions that are never really reconciled.	Aquarius will happily let Leo be the boss of the home. Some strange *objets d'art* will appear, but why not?	Both are interested in the other's mindset, but that isn't enough to make a home. Watery Pisces douses Leo's flame.
A pretty good match as Libra loves to cook dainty dishes and Virgo loves to eat them!	Scorpio will treat Virgo like a servant if allowed to get away with it. Virgo will rebel.	Virgo will fuss around Sagittarius and make demands; Sagittarius will try to escape.	Both are fussy homemakers, but they should be pretty much in tune.	No real conflict here, as Aquarius leaves Virgo to run the home in peace.	No particular problems here as long as each has some space for their own special interests.

	ARIES	TAURUS	GEMINI	CANCER	LEO	VIRGO
LIBRA WITH...	This will work as long as Libra is allowed to run the home and Aries is not too untidy.	No problem here, as both love cooking, gardening and making a beautiful home.	This works, because Libra does the housework and Gemini provides the entertainment.	Both do their share of the chores, but there may be disagreements on the way these are to be done.	Libra is more likely to do the chores and cook the meals, Leo will provide the cash.	Not a bad combination as both are tidy and will cooperate over the chores and childcare.
SCORPIO WITH...	Both love craft markets, boot and garage sales, and once they have acquired something they never get rid of it!	This rather depends upon the type of Scorpio that is involved, as some are home-makers but many are not really keen on home life.	Neither are interested in the home, so they will get someone in to do the housework while they both take the children away for the weekend.	Both like to relax at home and not bother too much with fussiness, and both are hoarders.	Both love being with their children. Leo will find plenty of jobs for Scorpio to do around the home. They may argue over money.	Both collect plenty of junk, but Virgo keeps their collection in some kind of order while Scorpio just collects it.
SAGITTARIUS WITH...	Both love to collect vast amounts of stuff. So as long as they don't argue over whose junk collection is the greater, they get on fine.	Sagittarius loves home life as long as they don't have to be there too much of the time, so Taurus will make most of the domestic decisions.	Gemini must live in a neat home and, while Sagittarius is not bothered, there is unlikely to be much disagreement here.	Having a settled base with family nearby is essential to Cancer, while Sagittarius feels stifled by this situation.	Leo wants to be the boss in the home and, while Sagittarius can adapt to this to some extent, they will eventually rebel.	Both have adaptable natures, although Virgo may be a little too fussy for Sagittarius's comfort.
CAPRICORN WITH...	This household could become a battleground, as both love to have their own way and make the major decisions.	These two can potter around together in the home and garden in perfect harmony.	Both are slightly fussy about the home and garden and both love to have family or friends in the home.	These two are in complete accord about domestic arrangements and family matters and both love to argue over petty things.	Capricorn's tendency to fuss over minor matters will drive Leo crazy, while Leo's irritability will get Capricorn down.	As long as Capricorn resists the urge to use Virgo as an unpaid servant, this can work well. Both enjoy a good moan about others.
AQUARIUS WITH...	Both are pretty laidback about housework and both love to have company drop in on them.	Taurus is more interested in the trappings of the home and in possessions than Aquarius, so Taurus will make the decisions.	Neither is fond of housework and both like to go out a lot. A cleaner would keep Gemini happy.	Both will try to run the home. As they have very different ideas, they will clash.	Aquarians are very understanding of Leo's dislike of housework and will act accordingly.	This can work, although they both need plenty of bookshelves and bandwidth for their vast collections of music and film.
PISCES WITH...	Aries will try to dominate Pisces and, while this will appear to work for a while, it won't work out in the long run.	Both love home and family and both enjoy pottering around the house and garden.	Pisces had better watch out if they don't wish to be left to do all the chores.	Both are terribly moody, but as long as they can understand each other, they can probably put up with each other's down times.	Both signs have a pretty easygoing attitude to home life and both like a comfortable home.	Each tries to serve the other and neither wishes to take the other for granted. They could fight about who *doesn't* do what!

LIBRA	SCORPIO	SAGITTARIUS	CAPRICORN	AQUARIUS	PISCES
Both love domestic life, cooking and playing with the children, and both know the value of relaxing in front of the TV.	Scorpio may not be as interested in homemaking as Libra, which is fine as long as Libra is happy to do it alone.	Libra will spend plenty of time alone in the home while Sagittarius is out and about doing their own thing.	These two will happily potter around the home like an old married couple, even if they are both teenagers.	Aquarius may be so house-proud that it hurts, or completely uninterested. The first suits Libra; the second is a dead loss.	Pisces needs one room to turn into a shrine, a temple, a healing centre, an artist's studio. Libra is quite happy with this.
Both love family life and both enjoy entertaining others in the home.	Either a major battleground over who does what, or a reasonable amount of cooperation and harmony. Each case is different.	Both can switch off from domestic matters, leave the dishes unwashed and take the children out for fun.	Neither is really bothered about the appearance of the home as long as there is plenty of good food in the fridge. Both love to go out a lot.	Either can be a fussy homemaker or a complete slob. As long as both are the same, peace will reign.	There is no real difficulty here, as long as the home is big enough for them to keep all their esoteric junk in.
Libra wants Sagittarius around at all times, but Sagittarius will find a job (or an excuse) that takes them away from the family.	Scorpio needs a home base and a place for the family to come back to, but neither makes their home the object of their attention.	As with all same-sign partnerships, Sagittarians rarely disagree on how things are done in the home.	Both want a large home in a nice location and neither is too bothered about tidiness.	This depends a bit on the Aquarian; if they are the laidback type, there will be no problem here.	There shouldn't really be a problem here, as these two have much the same outlook on home and family life.
Libra will work out what is to be done at home and Capricorn will do it – if they are in the mood!	No real problem here, as Capricorn will do what is necessary in the home while Scorpios do what they feel like.	Capricorn needs safety, security and a nice, regular routine. Sagittarius needs change and freedom.	Constant visits to and from both their families will keep them happily occupied in and around the home.	Capricorn will accuse Aquarius of spending too much time on non-essential things... but this will not matter in the least.	As long as Capricorn doesn't seek to push the Piscean around or be too demanding, this should be all right.
These two can live together as long as the Libran is not too argumentative, otherwise the home becomes a battleground.	Although both have a similar outlook on domestic matters, Scorpio's habit of criticizing can get Aquarius down.	As long as Aquarius doesn't expect Sagittarius to be at home much, this will be fine.	Both can be a bit fussy about decor and furnishings, but, if their tastes are similar, there should be no problem.	Aquarius can be very fussy or a real mess, so, if one kind pairs up with the other, trouble will break out.	The practicalities of running a home and a family will cause no difficulties, but they may find each other perplexing to live with.
Libra will try to dominate Pisces, which may work for a while, but Pisces will eventually get fed up.	No real problem here as neither is terribly house-proud, but both are moody and easily upset by bad atmospheres in the home.	Neither likes being tied to the home. Both will work out a way of doing the necessary chores so that they can go out frequently.	There is no real difference in the way these two wish to run their home and family, and both will also enjoy having a dog to care for.	This is either a haven of peace or a real battleground. Problems arise if Pisces tries to boss Aquarius around too much.	Both love a household full of dogs, small children and mess that defies description!

ADVENTURE COMPATIBILITY

	ARIES	TAURUS	GEMINI	CANCER	LEO	VIRGO
ARIES WITH...	Great; you'll share similar, energetic interests, and understand each other's needs.	These two share a love of music, perhaps also art, but little else mutually. Not enough.	Both enjoy gentle sport, but Gemini will avoid Aries's wilder schemes, so that's fine.	A shared love of family outings. Aries will encourage Cancer. Is that enough, though?	Both love fun and doing things on impulse. This relation-ship could be great fun.	No shared interests here, really. These two won't agree on anything for long.
TAURUS WITH...	A shared love of travel, but there's little else for these two to share. What a pity.	Both love to travel, but things could get stale unless they vary their ventures once in a while.	Artistic and musical interests appeal to both. Gemini will initiate and stimulate.	Not much adventure here, but these two won't mind; they will share many interests.	Both like family outings, and the children could be a common interest as well.	Taurus's lack of spontaneity may get Virgo down after a while. As long as that is handled, fine.
GEMINI WITH...	Get the boat, plane or skis ready for fun! These two will encourage each other.	Taurus will make Gemini a packed lunch... and stay behind. Their inclinations differ.	Both love walking, dancing and gentle sports. Here's a natural pair – great to see one like this.	Gemini has to encourage Cancer to try new things, and to participate more actively.	Both love to go out, but Leo won't share Gemini's active interests. Compromises can be found.	Holidays don't work too well. Gemini wants the sun and Virgo doesn't.
CANCER WITH...	Worlds apart, these don't have much in common. Aries is too impulsive for comfort.	A reasonable match. Both will enjoy similar pastimes, with less emphasis on the adventure.	Gemini is more inquisitive, which won't appeal. Why can't we go to the same place again and again?	Well-matched, you two! Whatever each does won't be far off the partner's mark.	Tastes differ here, but passably. Leo is more spontaneous and should choose the venue.	They'll help each other in whatever they do, and both will come back for more.
LEO WITH...	They can have so much spontaneous fun, they forget about their egos! Well, nearly...	Compromise is neither sign's strong point, so plenty of conflict is guaranteed.	Spontaneity is what it's all about, and willingness to try something new. No problem in this area!	The real adventure is watching each of these two trying to get away from the other's plans!	Excellent playmates, with the same tastes and dislikes. Lots of fun, but competitive games lead to bruises.	These two will get on well and understand each other's needs. At times, Virgo may resent Leo's limelight.
VIRGO WITH...	Aries loves dangerous and competitive sports; Virgo gets excited when exploring historical sites. So no mutual adventures.	Travel in considerable comfort suits both these signs and a sea cruise is enough adventure for both.	Adventure as such doesn't appeal to either of these nervous folk. However, visits to intriguing places delight both.	Neither is really adventurous, but Cancer can be quite footloose and will encourage Virgo to take off for the weekend.	Leo is fond of water sports and dancing. Virgo can be persuaded to dance but, while Leo swims, Virgo reads a book.	Neither is very adventurous, but they enjoy a game of pool or badminton on occasion.

LIBRA	SCORPIO	SAGITTARIUS	CAPRICORN	AQUARIUS	PISCES
Libra chooses the holidays, and sets the pace. Aries will escape to be with friends.	Both love to compete and to play hard. Fine, as long as it doesn't get serious.	Terrific togetherness, as both will take chances and push to the limit in whatever they do.	No shared interests here; Capricorn isn't sporty enough to keep up with Aries.	Aries will be surprised by Aquarius's hidden talents and adventurous streak.	No shared interests, no adventure, no fun for either of them. Their views are too different.
Similar interests, and neither likes too much excitement, so they should have fun together.	Scorpio will be frustrated by Taurus's apparent laziness; Taurus won't understand this.	Taurus will look after the children while Sagittarius goes out to play. That's the only way this will work.	The greatest excitement comes from finding a new tax-saving scheme! And that's fine!	Taurus can't see why Aquarius wants to go out to so many meetings all the time.	Shared interests of a quiet nature. Good harmony and understanding of each other's tastes.
Both love to shop and eat out. There's adventure in that, too!	It's best if they have separate interests. That can be a good solution for many sign combinations.	A good match – not only in tennis! These two can keep each other going very well.	Travelling suits both, ideally to restful places. Change destinations each time, though.	Aquarius's crazy ideas go down well with Gemini. Never a dull moment for either of them!	Only intellectual interests in common, really. Pisces won't want separate holidays, either.
Libra will organize the outing, Cancer will pack a nice lunch. The youngsters will be happy too.	Yes, the competitiveness is a drawback, but Cancer is ready for a game now and again.	No and no again. Sagittarius will be restless within half an hour, most of the time.	Not bad at all. They should develop and concentrate on some shared interests.	The wise Cancerian will let Aquarius do their thing alone, mostly. That'll save on tranquillisers!	They'll trundle around the countryside quite happily together. Good company for each other.
Well-matched, these two. Leo will lead. Libra will, too, but by diplomacy!	With the best of intentions, things eventually become competitive and frustrating for one or both.	Well-starred. Both enjoy themselves and are ready for spontaneous pursuits.	Works only if, by chance, both have an identical interest. That's rare; they like different playgrounds.	Funnily enough, the opposite sexes will share a lot of pursuits very happily. Not so for the same sex.	Not much in common, though not a competitive problem. Each spells 'adventure' differently!
Any event in an area that has a military or historical significance will be attended by both.	Scorpio loves danger and excitement; Virgo hates it, so they will have separate interests.	Sagittarius is the most adventurous sign in the zodiac, Virgo is nervous of adventure.	Capricorn will take Virgo travelling and both need comfort and security while on the move.	Aquarius is quite inventive and mildly sporty and this fits well with Virgo's temperament.	Pisces might frighten the wits out of Virgo by suggesting that they go water-skiing or parasailing.

	ARIES	TAURUS	GEMINI	CANCER	LEO	VIRGO
LIBRA WITH...	Aries will encourage Libra to explore the competitive side of their nature.	Much the same attitude, as both see a gym just as a place to dress in all the latest gear.	Gentle sport, enjoyable holidays and visiting friends appeal to both.	Both love history, so holidays that take in ancient or historical places will suit both parties.	Both enjoy non-contact sports, such as golf and swimming, and both love dancing, parties and celebrations, but neither likes danger.	These two have similar tastes, but Libra may be more of a party animal than workaholic Virgo.
SCORPIO WITH...	No problem here as both love sport, games and travel to unusual locations. Life must be full of change and adventure for both parties.	These two are absolute opposites as Taurus needs a quiet life, but they share an interest in quiet matters, such as music and books.	Gemini will just have to leave Scorpio to ski and run and then join in when Scorpio decides to play badminton or pool.	Little mutual interest here, except perhaps for swimming and water sports, although both love holidays on cruise ships.	Both are happy to do mad things on the spur of the moment, so no real problem here.	Intellectual pursuits can be shared and both love music, drama and entertainment, but Virgo doesn't go in for sport.
SAGITTARIUS WITH...	Aries and Sagittarius are both adventurous and athletic types, so they will enjoy doing lots of things together.	These two are on different planets as far as adventurous pursuits are concerned, so each will have to do their own thing.	Gemini is far too nervous to do anything risky, so Sagittarius will have to go hang-gliding without their partner.	Both love to travel, although Cancer needs to do so in comfort while Sagittarius can rough it. Sport is not likely to become a shared interest.	Although both are fairly adventurous, their interests are not exactly the same so these will have to diverge a little.	Virgo's idea of adventure is reading an exciting book or going to a good film. Sagittarius soon gets bored with this limited outlook.
CAPRICORN WITH...	There could be a reasonable amount of similarity here, as long as Capricorn is the quieter type.	Both enjoy having fun on holiday and neither is into contact sports, so there is no conflict here.	No real problem here, as Gemini is likely to encourage Capricorn to have a go at something new from time to time.	As long as the family is included in anything that they do, both are happy with any shared adventures.	This is an area where these two can be quite happy together, as neither likes dirty, wet or dangerous activities.	Neither is particularly adventurous, but Capricorn will set the pace and Virgo will follow.
AQUARIUS WITH...	Either can be adventurous or more interested in politics. If they are of the same mind, then this is great.	This depends on the unpredictable Aquarian, as they may have quirky interests that Taurus just doesn't share.	If there is money to spare, they will travel. If not, both will read and enjoy sport together.	Cancer's idea of adventure is trying out a new recipe, Aquarius's is trying out paragliding.	Absolutely no problem here. Both love sport and outdoor activities when time (and money) permit.	Aquarius is easily bored, so will get Virgo out to do things. Virgo probably won't mind this too much.
PISCES WITH...	Both can enjoy swimming and both love doing things with the family, so this could be a success.	Pisces is a bit more adventurous than Taurus, but there is no real conflict here.	Both love to go out, but their interests don't really coincide. Separate outings may be the answer.	Cancer will set the pace, but Pisces doesn't mind. Both love to do things with the family in tow.	Anything that includes the whole family pleases both these signs. They enjoy much the same things.	Pisces is slightly more adventurous than nervous Virgo, but Virgo can be persuaded to try a few new things.

LIBRA	SCORPIO	SAGITTARIUS	CAPRICORN	AQUARIUS	PISCES
Both need comfort more than adventure, but both enjoy travel and outings to beautiful places and lovely gardens.	Scorpio loves dangerous sports and pastimes, while Libra prefers to lounge by the pool.	Sagittarius will try to introduce Libra to hang-gliding, surfing and being shot from a cannon; Libra will quail in horror at it all!	Both might fancy going on safari... but only in comfort and with the animals at a nice safe distance!	Both enjoy gentle sports, gambling and spontaneous outings. Neither is truly adventurous, but both will try most things once.	Pisces is more spontaneous than Libra, who needs to pack even for a short walk.
Apart from travel and music, there is little in common between these two. Libra is too laid back for energetic sport or other taxing activities.	Both need plenty of exercise, so joint membership of the local gym is a good idea.	Both love sport, travel and meeting new and interesting people.	Both enjoy exotic travel and anything to do with animals, so a safari or horse riding might be a good joint pastime.	Scorpio is more competitive and more interested in danger than Aquarius, but both love to try something new.	Both love travel and interesting experiences and both love sport, so no conflict here.
Libra is not an adventurous sign – except perhaps for business matters – so there's little meeting of minds here.	Anything Sagittarius dreams up, Scorpio will try. So as long as they enjoy being team mates, this will be fine.	Let us hope that these two can stagger their sports injuries so that they are not both laid up at the same time!	Apart from travel and holidays, there is not much in the way of shared interests here.	If both are keen on the same thing, this is fine. There is no guarantee, as both are quirky and have their own interests.	Pisces loves to experience things and the stranger these are, the better. Sagittarius will happily join in.
Libra is more spontaneous and adventurous than Capricorn, but this is not a major problem.	Scorpio will find Capricorn to be something of a stick-in-the-mud, but they do share the same kind of holiday interests.	There is very little for these to share, except perhaps for intellectual interests.	These two should book a nice cruise around the Caribbean, as they both will love it.	Both are fascinated by new ideas and both love to read, think and talk things over. A happy combination.	Pisces wants to go off on slightly crazy escapades, while Capricorn needs notice of even a mildly adventurous outing.
Both enjoy gentle sports; they may encourage each other to take up flying.	Both are adventurous and they share many intellectual interests, so this can work well.	Both can be sporty or they can be keen on philosophy and intellectual pursuits, so this can be great.	Capricorn won't really want to get up and go when Aquarius feels like it, so conflict can arise.	Both enjoy gentle sport and neither is particularly competitive, so no conflict here.	Shared interests are very likely, especially going to spiritualist meetings, and some outdoor pursuits.
Both enjoy travel and outings and both love to have friends and family around to visit, but neither is truly adventurous.	Scorpio can encourage Pisces to try exciting and risky stunts. For the most part Pisces will give this a go.	There is a fair chance that these two will share interests in esoteric subjects, such as astrology, and also sporting pursuits.	Neither may be over-interested in sport, but both love to sing and dance so anything that indulges these interests would suit them.	These two are more likely to share intellectual interests than to go looking for adventure.	No real conflict here as both are quite adventurous and both love doing new things on the spur of the moment.

PASSION COMPATIBILITY

	ARIES	TAURUS	GEMINI	CANCER	LEO	VIRGO
ARIES WITH...	Who'll be the boss? That's the question here! The rest is spontaneous and very active.	Aries may be irritated by Taurus's slow pace, while Taurus will feel uncomfortably rushed.	This is a good combination, as both warm up very quickly. They will both enjoy the relationship.	A surprisingly good combination, as long as Cancer feels loved and wanted.	Hot, hot, hot! Aries will start, and Leo won't want to stop! What more can you ask for?	About enough spark to start things up, but not for long. A fling, yes; love, no.
TAURUS WITH...	Great while love is in the air, but not long term. Aries's dynamo is too much.	Wonderful! The love and passion will never die, these two will get on very well.	Very good, as Gemini cherishes Taurus's reliability and attention in love.	Both appreciate comfortable, relaxed love-making, and the important thing to them is mutual happiness.	This works well, as both need stability and warmth in a relationship, which is readily forthcoming.	Both are sensual, and Taurus appreciates Virgo's enthusiasm! Well-matched, all told.
GEMINI WITH...	Talk, laughs and great sex will make this team a winner! Fun all round.	A good combination, as both share strong feelings. Some variety can spice things up.	Frequent sex keeps both happy. Not only apples each day keep the doctor away!	Cancer is likely to kill Gemini's passion by saying the wrong thing at the wrong time.	Fun, laughs, game-playing, then more fun, laughs and game-playing... it's a good relationship.	A surprisingly good combination, as both are spontaneous and enjoy each other.
CANCER WITH...	Aries has a huge appetite for love and passion, so an affair with this sign can be exhausting!	Each will enjoy the other's tenderness, and savour the afterglow. An extremely satisfactory situation.	The pace is a bit uneven, and Cancer may put his or her foot in it at times, spoiling the mood.	No dizzy heights of passion, but who needs that anyway? The closeness is the thing.	Very similar to Aries in this department! Where does that appetite come from?	These two can heal each other's heartaches and other pains. Cancer, just don't cling too much.
LEO WITH...	Sparks fly in this relationship, so when you aren't arguing you can have a lot of fun together.	There have been better matches made in heaven, for sure. Neither wants to change their style.	A good combination, and Gemini will adore the warmth and enthusiasm.	Not a disaster, but one or the other will feel unfulfilled. The inner needs are very different.	This reaches the heights, although each will bruise the other. But they don't hold grudges for long...	A strange combination. No roaring fires here, yet an instinctive, mutual love can flourish.
VIRGO WITH...	Virgo is more earthy and passionate than one might think, so a short-term fling could be fun.	Plenty to enjoy here as both signs are sensual, but Virgo's coolness after sex might cause Taurus anxiety.	Both are extremely sexy when in the mood, but on which dates do their moods coincide?	If Virgo is in the mood to be mothered, Cancer is the one to do it. Otherwise, Virgo could end up feeling smothered.	Lovemaking is no problem, but Leo is affectionate and tactile, while Virgo keeps a distance before and after sex.	They have the same needs, but both might want to take the lead, so they will have to take turns.

LIBRA	SCORPIO	SAGITTARIUS	CAPRICORN	AQUARIUS	PISCES
Both are highly sexed, but Libra won't keep up with Aries. Not a long-term arrangement.	No problem here, but each must have their turn at being in control, then both will be happy.	Aries may meet his or her match here. Wow, this could be an interesting partnership!	When Aries becomes really ardent, Capricorn will make a break for freedom! It's not easy to keep up.	This works well, but talk will be as important as sex to Aquarius. That could be a damper.	Short, sharp sex perhaps, but no shared wavelength to keep up the momentum.
This couple really score well here! Each easily gathers the other's likes and dislikes.	This can be OK, as long as Taurus is prepared to try something new.	A short-term fling is fine for Sagittarius. Taurus won't like the brevity and will be hurt.	Passion? What's that? Capricorn would need to make an effort at this relationship.	This is about the only area where there might be an ongoing attraction. It's worth a try.	Gentle loving makes both happy. Each sign would appreciate the other's style.
Good for both, even if they are not in love with each other. A good match.	A fairly good outlook for a fling, but no more. Well, that's better than nothing.	Not a long-term proposition, but it's good while it lasts. Both would appreciate that.	Neither excites the other very much. Unfortunately they can't find the same wavelength.	Languid love-making and long talks please both. No problem here, except for loss of sleep!	Pisces needs more consistency than Gemini can give. It becomes frustrating for both.
How nice of Libra to bring that small gift. And my headache isn't that bad after all...	Both will relish the emotional involvement. This will work well with Scorpio leading the way.	Hardly a long-term involvement, as Cancer soon discovers. How did it get this far, anyway?	Cancer will feel safe and warm, perhaps loosening up Capricorn's suspicions in the process.	Cancer, I told you at least once this year that I love you! What more do you want?	Pisces will soothe Cancer's nerves, and they will both feel uplifted. Well and good.
Jealousy can easily rear its head if Libra doesn't let Leo have their share of the limelight.	These two fixed people will either form an intense bond or totally reject each other. There's no in-between.	A genuine, friendly relationship will exist here. They'll give each other space and appreciation.	Much effort and understanding are needed here, and some time must pass before deep feelings arise.	Lots of talk, lots of laughs, real feelings in between. Allow the other their own identity, and it's great!	This can be fascinating if both are reasonable people. But can you keep it up?
No problem here, but Virgo may need more frequent sex than Libra.	An affair of the heart will be very intense, but a long-term relationship? Perhaps not.	This can work for a while, but Sagittarius's 'anything goes' attitude makes Virgo feel insecure.	Love runs deep between these two and sex works well, as both improve with age!	Either terrific or a complete waste of time. Emotionally, both can be cool and unrevealing.	Not a bad passion quotient, but Virgo may be a little cool for Pisces, while Pisces may be a little too playful for Virgo.

	ARIES	TAURUS	GEMINI	CANCER	LEO	VIRGO
LIBRA WITH...	A pretty good passion quotient, although Aries falls deeply in love while Libra holds back a little.	Both are sensual and loving, but neither likes to show their feelings in public.	Not a bad combination, as long as both remain faithful and don't fool around with others.	This area of life is fine, as long as Cancer gets the affection they need.	A sizzling combination as both love to make love, but Libra must be careful not to turn off too quickly once the deed has been done.	Both are easily tempted into leaving the chores when they could be sharing a sensual bath together!
SCORPIO WITH...	Terrific sex and terrific fights and arguments, so if they don't wear each other out in bed, they might just kill one another while out of it.	This might work for a while as both are sensual, but Scorpio will want more adventure in bed than Taurus can be bothered with.	This is fine as both enjoy sex, but may become irritated by each other's self-absorption.	Cancer gives the love and comfort that Scorpio needs, but both are jealous, so they may accuse each other of unfaithfulness.	Scorpio needs more sex than Leo, but despite this Leo may be unfaithful. Confusion causes arguments.	Both are sensual and passionate. Virgo wants to talk when Scorpio wants to be loving and quiet, so a little friction can ensue.
SAGITTARIUS WITH...	No problem here, as both are spontaneous and love to experiment. They both also have hot tempers, so fights will occur.	For a short-lived love affair, passionate sex will keep things alive, but over a longer time span it is unlikely to be enough.	These two meet mentally and physically, as both relieve their tensions with sex.	This can work but, unless Cancer is given reassurance, passionate feelings will quickly die away and Sagittarius will become bored.	Both are highly sexed and extremely passionate, so this could be the glue that holds this one together.	Both are passionate, but Sagittarius is more experimental than Virgo. Disagreements will soon set in.
CAPRICORN WITH...	Capricorn can be passionate, but sex comes low on the list of their priorities. Little in common here.	Both are sensual but not especially adventurous. No difficulty in this area of life either.	Oddly enough, this is likely to be a fairly peaceful area of life for these two, so success in this area as well.	No real problem here as both like to take things easy and enjoy the time they spend together in loving harmony.	This is probably best left until later in life, as a young Leo will want more than Capricorn is prepared to give.	The relationship could become slightly unstuck here, as Capricorn will seek to dominate the situation.
AQUARIUS WITH...	Aries might like more variety in bed than Aquarius, but Aquarius can usually be persuaded.	Both are sensual and passionate and neither is critical of the other's performance.	These two signs have a fondness for some experimentation and a willingness to communicate clearly to each other.	As long as the mobile is switched off so that Cancer's family can't interrupt them, this can be quite good.	Leo is very affectionate; some Aquarians are and some are not. It depends on the Aquarian in question.	Not a bad meeting of minds and bodies here, as both are willing to experiment.
PISCES WITH...	Both are passionate, but Pisces needs more love and affection than Aries may wish to give.	Quite a good combination, as both are sensual and neither is bossy where sex is concerned.	Both love to make love, although Pisces will be a touch more adventurous than Gemini.	A shared laugh or two, a good meal and a nice glass of wine and these two can really get it together.	This can work as long as the rest of the relationship is in good shape. Otherwise, moodiness and arguments can creep in.	This works well, because they will take the trouble to make each other happy in bed.

LIBRA	SCORPIO	SAGITTARIUS	CAPRICORN	AQUARIUS	PISCES
This is where these two become really adventurous, taking time off for love at any hour of the day or night.	Not a bad combination for a really high-octane fling, but living together will cool the ardour of both.	Both are prepared to try anything once, and often many more times. Both can indulge their fantasies to the hilt.	Great... until Capricorn is asked to do something different, at which time they develop a sudden, very convenient ailment.	Terrific, wonderful; these two amazingly sexy signs really meet their match in each other.	There may be a mutual attraction between these two signs, but this could soon be replaced by irritation.
This is an area of great success for this pair, as both love to get down to business at the drop of a hat.	Both relieve their considerable tensions with frequent sex.	Sagittarius will try anything feasible and Scorpio will encourage Sagittarius to do so.	Capricorn can be passionate, but not always about sex, and they may have half an ear out for their mobile phone.	This is either a wonderful experience for both or a complete disaster. Both have an all-or-nothing attitude.	Both are moody and apt to be spontaneous about sex, so as long as their moods coincide this can work well.
Not a bad combination, but it is hard to work out who will take the lead or set the pace.	If Scorpio's moods and Sagittarius's sarcasm can be put aside for a while, this is great.	Goodness only knows what these two will get up to... but some of their furniture will get broken in the process.	Sagittarius will suggest doing something strange and Capricorn will pretend to have a bad back, thus avoiding the action.	No real problem here, although Aquarius likes to set the pace, which may pall for Sagittarius after a while.	The chances are that both will move on sooner or later but, in the short term, the passion stakes will be high.
This could work, as both are sensual, but there may not be much of a spark between them.	Unless there is something else going for this relationship, Scorpio will soon get bored.	After a few rather messy encounters, these two will give up and go their own ways.	Great stuff – neither sign is fond of anything too outrageous, so both are happy.	Both get better as they get older, so, if they stay together, they will play together.	Pisces is into fun, while Capricorn takes everything seriously, so this is a potential problem.
No problem, as both love to leave the chores in favour of a half-day's passionate love-making.	Even after a separation these signs can get together for fairly regular sex, as it is a mutual interest.	Sagittarius might be a little too demanding for Aquarius, but this shouldn't be much of a problem.	This can be a reasonable area of agreement and, if so, it gets better as time goes by.	Aquarians are extremely passionate, but they can also be cool and logical. Let's hope both feel the same way at the same time.	There shouldn't be too much of a problem here, as long as neither of them drinks too much!
No problem here, although Pisces might get a bit bored if Libra refuses to alter their usual routine.	Both are quite passionate, but the moodiness of both signs could cause problems.	Both like a bit of a change now and then, so a visit to the sex shop should give them both a good time.	Capricornians can be very passionate when they are in love, but Pisces can blow hot and cold depending on the mood.	Both are very passionate at times, but some Aquarians are too detached for Pisces and some Pisceans too moody for Aquarius.	If both want to be the boss they won't stay together long.

This edition first published in the United Kingdom in 2021 by
Collins & Brown
43 Great Ormond Street
London
WC1N 3HZ

An imprint of Pavilion Books Company Ltd

Distributed in the United States and Canada by
Sterling Publishing Co., Inc. 1166 Avenue of the Americas, New York, NY 10036

ISBN 978-1-911163-92-3

A CIP catalogue record for this book is available from the British Library.

10 9 8 7 6 5 4 3 2 1

Reproduction by Mission, Hong Kong
Printed by 1010 Printing International Ltd, China

Publisher: Helen Lewis
Editor: Izzy Holton
Designer: Alice Kennedy-Owen
Production Controller: Jessica Arvidsson
Copy Editor: Lucy Bannell
Illustrator: Amy Blackwell

www.pavilionbooks.com